NONEQUILIBRIUM STATISTICAL MECHANICS

NONEQUILIBRIUM STATISTICAL MECHANICS

Robert Zwanzig

UNIVERSITY PRESS

2001

OXFORD
UNIVERSITY PRESS

Oxford   New York
Athens   Auckland   Bangkok   Bogotá   Buenos Aires   Cape Town
Chennai   Dar es Salaam   Delhi   Florence   Hong Kong   Istanbul   Karachi
Kolkata   Kuala Lumpur   Madrid   Melbourne   Mexico City   Mumbai   Nairobi
Paris   São Paulo   Shanghai   Singapore   Taipei   Tokyo   Toronto   Warsaw

and associated companies in
Berlin   Ibadan

Copyright © 2001 by Oxford University Press

Published by Oxford University Press, Inc.
198 Madison Avenue, New York, New York 10016

Oxford is a registered trademark of Oxford University Press

All rights reserved. No part of this publication may be reproduced,
stored in a retrieval system, or transmitted, in any form or by any means,
electronic, mechanical, photocopying, recording or otherwise,
without the prior permission of Oxford University Press.

Library of Congress Cataloging-in-Publication Data
Zwanzig, Robert.
   Nonequilibrium statistical mechanics / Robert Zwanzing.
      p.   cm.
   ISBN 0-19-514018-4
   1. Nonequilibrium statistical mechanics.   I. Title.
QC174.86.N65 Z93   2000
530.13—dc21   00-023880

9 8 7 6 5 4 3 2

Printed in the United States of America
on acid-free paper

Preface

My intention in this book is to provide some understanding of the various methods that have been proposed for treating time-dependent processes in statistical mechanics. Some familiarity with equilibrium statistical mechanics is assumed. Whenever possible, I start with a simple example, generalize it, and finally discuss its theoretical foundations. The applications treated here were chosen as simple illustrations of a particular method; these choices were motivated by their utility in chemical physics. The methods, of course, have a much wider applicability, for example, in biophysics or condensed matter physics or even in astrophysics. There are no problem sets or exercises; most interesting problems are suitable subjects for serious research, and there is no point practicing on uninteresting problems. There are few literature references, only the occasional name and date; a lot of the material has been around a long time, and some of it is my own work.

In a letter accepting the Rumford medal of the American Academy of Arts and Sciences in 1881, J. Willard Gibbs wrote

> One of the principal objects of theoretical research in any department of knowledge is to find the point of view from which the subject appears in its greatest simplicity.

This is a hard standard; I hope that I came close. I am especially indebted to Attila Szabo, who encouraged me to finish a project I started about 1965 and who worked hard to get me to simplify my often obscure treatments of various topics. My failures are my own, not his.

Contents

1. Brownian Motion and Langevin Equations 3

    1.1 Langevin Equation and the Fluctuation-Dissipation Theorem 3
    1.2 Time Correlation Functions 7
    1.3 Correlation Functions and Brownian Motion 10
    1.4 Brownian Motion of Other Variables 14
    1.5 Generalizations of Langevin Equations 18
    1.6 Brownian Motion in a Harmonic Oscillator Heat Bath 21
    1.7 Heavy Mass in a Harmonic Lattice 24

2. Fokker-Planck Equations 30

    2.1 Liouville Equation in Classical Mechanics 30
    2.2 Fokker-Planck Equations 36
    2.3 About Fokker-Planck Equations 41

3. Master Equations 48

    3.1 The Golden Rule 48
    3.2 Optical Absorption Coefficient 53

3.3 Quantum Mechanical Master Equations 56
3.4 Other Kinds of Master Equations 61

4. Reaction Rates 67

4.1 Transition State Theory 67
4.2 The Kramers Problem and First Passage Times 73
4.3 The Kramers Problem and Energy Diffusion 78

5. Kinetic Models 83

5.1 Kinetic Models 83
5.2 Kinetic Models and Rotational Relaxation 89
5.3 BGK Equation and the H-Theorem 93
5.4 BGK Equation and Hydrodynamics 96

6. Quantum Dynamics 101

6.1 The Quantum Liouville Operator 101
6.2 Electron Transfer Kinetics 106
6.3 Two-Level System in a Heat Bath: Dephasing 110
6.4 Two-Level System in a Heat Bath: Bloch Equations 115
6.5 Master Equation Revisited 121

7. Linear Response Theory 127

7.1 Static Linear Response 127
7.2 Dynamic Linear Response 130
7.3 Applications of Linear Response Theory 136

8. Projection Operators 143

8.1 Projection Operators and Hilbert Space 143
8.2 Derivation of Generalized Langevin Equations 149
8.3 Noise in Generalized Langevin Equations 151
8.4 Generalized Langevin Equations—Some Identities 157
8.5 From Nonlinear to Linear—An Example 160
8.6 Linear Langevin Equations for Slow Variables 165

9. Nonlinear Problems 169

   9.1 Mode-Coupling Theory and Long Time Tails 169
   9.2 Derivation of Nonlinear Langevin Equations and Fokker-Planck Equations 174
   9.3 Nonlinear Langevin Equations and Fokker-Planck Equations for Slow Variables 181
   9.4 Kinds of Nonlinearity 185
   9.5 Nonlinear Transport Equations 188

10. The Paradoxes of Irreversibility 193

Appendixes 198

   1 First-Order Linear Differential Equations 198
   2 Gaussian Random Variables 200
   3 Laplace Transforms 203
   4 Continued Fractions 205
   5 Phenomenological Transport Equations 207

References 211

Index 213

NONEQUILIBRIUM STATISTICAL MECHANICS

1

Brownian Motion and Langevin Equations

1.1 Langevin Equation and the Fluctuation-Dissipation Theorem

The theory of Brownian motion is perhaps the simplest approximate way to treat the dynamics of nonequilibrium systems. The fundamental equation is called the Langevin equation; it contains both frictional forces and random forces. The fluctuation-dissipation theorem relates these forces to each other. This theorem has many important and far-reaching generalizations. For the present, we focus on the most elementary version of the theorem.

The random motion of a small particle immersed in a fluid is called *Brownian motion*. Early investigations of this phenomenon were made on pollen grains, dust particles, and various other objects of colloidal size. Later it became clear that the theory of Brownian motion could be applied successfully to many other phenomena, for example, the motion of ions in water or the reorientation of dipolar molecules.

In particular, the theory of Brownian motion has been extended to situations where the "Brownian particle" is not a real particle at all, but instead some collective property of a macroscopic system. This might be, for example, the instantaneous concentration of any component of a chemically reacting system near thermal equilibrium. Here the irregular fluctuation in time of this concentration corresponds to the irregular motion of the dust particle. This kind of extension is of the greatest importance and will be discussed in depth later.

While the motion of a dust particle performing Brownian motion appears to be quite random, it must nevertheless be describable by the same equations of motion as is any other dynamical system. In classical mechanics, these are Newton's or Hamilton's equations.

Consider the one-dimensional motion of a spherical particle (radius a, mass m, position x, velocity v) in a fluid medium (viscosity η). Newton's equation of motion for the particle is

$$m\frac{dv}{dt} = F_{\text{total}}(t), \qquad (1.1)$$

where $F_{\text{total}}(t)$ is the total instantaneous force on the particle at time t. This force is due to the interaction of the Brownian particle with the surrounding medium. If the positions of the molecules in the surrounding medium are known as functions of time, then in principle, this force is a known function of time. In this sense, it is not a "random force" at all. An example that illustrates this point, a Brownian particle coupled to a heat bath of harmonic oscillators, will be discussed later.

It is usually not practical or even desirable to look for an exact expression for $F_{\text{total}}(t)$. Experience teaches us that in typical cases, this force is dominated by a frictional force $-\zeta v$, proportional to the velocity of the Brownian particle. The friction coefficient is given by Stokes' law, $\zeta = 6\pi\eta a$. If this is the whole story, the equation of motion for the Brownian particle becomes

$$m\frac{dv}{dt} \cong -\zeta v, \qquad (1.2)$$

and, as a linear first-order differential equation, it has the familiar solution

$$v(t) = e^{-\zeta t/m}v(0). \qquad (1.3)$$

According to this, the velocity of the Brownian particle is predicted to decay to zero at long time. This cannot be strictly true because the mean squared velocity of the particle at thermal equilibrium is $\langle v^2 \rangle_{\text{eq}} = KT/m$, so that the actual velocity cannot remain at zero. Evidently, the assumption that $F_{\text{total}}(t)$ is dominated by the frictional force must be modified.

The appropriate modification, suggested by the observed randomness of an individual trajectory, is to add a "random" or "fluctuating" force $\delta F(t)$ to the frictional force, so that the equation of motion becomes

$$m\frac{dv}{dt} = -\zeta v + \delta F(t). \qquad (1.4)$$

This is the *Langevin equation* for a Brownian particle. In effect, the total force has been partitioned into a systematic part (or friction) and a fluctuating part (or noise). Both friction and noise come from the interaction of the Brownian particle with its environment (called, for convenience, the "heat bath"). Because of this, one should not be surprised to find that there is a fundamental relation between friction and noise; this will be demonstrated shortly.

There are two basic views of the nature of the fluctuating force. In the more-commonly presented view, the fluctuating force is supposed to come from occasional impacts of the Brownian particle with molecules of the surrounding medium. The force during an impact is supposed to vary with extreme rapidity over the time of any observation, in fact, in any infinitesimal time interval. This clearly cannot be strictly true in any real system. Then the effects of the fluctuating force can be summarized by giving its first and second moments, as time averages over an infinitesimal time interval,

$$\langle \delta F(t) \rangle = 0, \qquad \langle \delta F(t) \delta F(t') \rangle = 2B\delta(t-t'). \qquad (1.5)$$

B is a measure of the strength of the fluctuating force. The delta function in time indicates that there is no correlation between impacts in any distinct time intervals dt and dt'. The remaining mathematical specification of this dynamical model is that the fluctuating force has a Gaussian distribution determined by these moments.

The other view can be illustrated by the analogy of random number generators in computers. These algorithms are deterministic; that is, if the same seed in used in repetitions of the algorithm, the same sequence of numbers is generated. Yet the sequence generated by a good algorithm is "random" in the sense that it satisfies various statistical requirements of randomness for almost all choices of seed. The output of a random number generator is used as input to other programs, for example, Monte Carlo integration. The results are generally independent of the initial seed; only the statistical distribution of random numbers is important. In the same way, the randomness of Brownian noise is fully determined by the initial state of the heat bath. The results of a calculation using the Langevin equation are expected to be independent of the initial state and to involve only the statistical distribution of the noise. In this view, the averages in eq. (1.5) come from averages over initial states. A later section shows how all this can come from a simple harmonic oscillator model of a Brownian heat bath.

As remarked earlier, the particle's velocity decays to zero in the absence of noise, but this cannot be so. At thermal equilibrium, we must require that $\langle v^2 \rangle_{eq} = kT/m$. The Langevin equation, which is a

6 NONEQUILIBRIUM STATISTICAL MECHANICS

linear, first-order, inhomogeneous differential equation, can be solved to give

$$v(t) = e^{-\zeta t/m}v(0) + \int_0^t dt' e^{-\zeta(t-t')/m}\,\delta F(t')/m. \tag{1.6}$$

(Appendix 1 deals with solutions of equations of this kind.) The first term gives the exponential decay of the initial velocity, and the second term gives the extra velocity produced by the random noise. Let us use this to get the mean squared velocity. There are three contributions to $v(t)^2$; the first one is

$$e^{-2\zeta t/m}v(0)^2 \tag{1.7}$$

and clearly decays to zero at long times. There are two cross terms, each first order in the noise,

$$2v(0)e^{-\zeta t/m}\int_0^t dt' e^{-\zeta(t-t')/m}\,\delta F(t')/m. \tag{1.8}$$

On averaging over noise, these cross terms vanish. The final term is second order in the noise:

$$\int_0^t dt' e^{-\zeta(t-t')/m}\delta F(t')\int_0^t dt'' e^{-\zeta(t-t'')/m}\,\delta F(t'')/m^2. \tag{1.9}$$

Now the product of two noise factors is averaged, according to eq. (1.5), and leads to

$$\int_0^t dt' e^{-\zeta(t-t')/m}\int_0^t dt'' e^{-\zeta(t-t'')/m}\,2B\delta(t'-t'')/m^2. \tag{1.10}$$

The delta function removes one time integration, and the other can be done directly. The resulting mean squared velocity is

$$\langle v(t)^2 \rangle = e^{-2\zeta t/m}v(0)^2 + \frac{B}{\zeta m}(1 - e^{-2\zeta t/m}). \tag{1.11}$$

In the long time limit, the exponentials drop out, and this quantity approaches $B/\zeta m$. But in the long time limit, the mean squared velocity must approach its equilibrium value kT/m. Consequently we find

$$B = \zeta kT. \tag{1.12}$$

This result is known as the *Fluctuation-dissipation theorem*. It relates the strength B of the random noise or fluctuating force to the magnitude ζ of the friction or dissipation. It expresses the balance between friction, which tends to drive any system to a completely "dead" state, and noise, which tends to keep the system "alive." This balance is required to have a thermal equilibrium state at long times. Many

BROWNIAN MOTION AND LANGEVIN EQUATIONS   7

variations on the fluctuation-dissipation theorem will be encountered in the following pages.

1.2 Time Correlation Functions

The Langevin equation can be used to calculate various time correlation functions. This section provides an introduction to these important quantities.

Equilibrium statistical mechanics is based on the idea of a statistical ensemble. We learn that the thermodynamic properties of a gas, for example, can be found by calculating the partition function of a statistical ensemble. We learn that the spatial structure of a liquid can be described statistically by a pair correlation function.

Nonequilibrium statistical mechanics is based on the same idea of a statistical ensemble. A fundamental difference, however, is that while there is only one equilibrium state, there are many nonequilibrium states. There is no unique "partition function" to use as a starting point for calculating transport properties. Time correlation functions play the same role as partition functions and spatial pair correlation functions in nonequilibrium statistical mechanics. Many properties of systems out of equilibrium, for example, coefficients of viscosity, thermal conductivity, diffusion, and conductivity, are determined by time correlation functions. They also provide a useful way to interpret experiments on neutron and light scattering, optical spectroscopy, and nuclear magnetic resonance.

We encounter a time correlation function whenever we analyze the statistical behavior of some time-dependent quantity $A(t)$ measured over a long time. The quantity $A(t)$ could be, for example, the intensity of light scattered by fluctuations in a liquid, or it could be the velocity of a single particle followed in a computer simulation of a liquid. The first stage in the analysis is to time-average the quantity itself,

$$\langle A \rangle = \frac{1}{\tau} \int_0^\tau dt A(t). \tag{1.13}$$

Then we subtract the average to get the fluctuation δA,

$$\delta A(t) = A(t) - \langle A \rangle. \tag{1.14}$$

One often observes that fluctuations at different times are correlated (in the same way that molecules in a liquid are spatially correlated). The time-averaged product of two fluctuations at different times,

$$C(t) = \frac{1}{\tau} \int_0^\tau ds \, \delta A(s) \, \delta A(t+s), \tag{1.15}$$

is called the *time correlation function* (TCF) of δA. The conventional mean squared fluctuation, the time average of fluctuations at the same time, is $C(0)$.

If the system under investigation is ergodic (generally assumed without proof), a long time average is equivalent to an equilibrium ensemble average. This is where the methods of statistical mechanics come in. Just as we get a pressure by calculating the partition function of a statistical ensemble instead of making a long time average of a single sample, we get a time correlation function by calculating an ensemble average of the product of two fluctuations instead of its long time average. In an equilibrium ensemble, there is no special initial time, and $C(t)$ depends only on the difference t between the two times.

While we based the definition of $C(t)$ on a record of the time dependence of $A(t)$, of the sort that might be produced, for example, by a computer simulation, many experiments actually generate the Fourier transform of the time correlation function directly. Generally, the Fourier transform of any time correlation function,

$$C_\omega = \int_{-\infty}^{\infty} dt\, e^{-i\omega t} C(t), \qquad (1.16)$$

is called its *spectral density*. If we know the spectral density, we can recover the time dependence of the correlation function by Fourier inversion. For example, the optical absorption spectrum of a system as a function of frequency is related to the time correlation function of its total electric dipole moment. This connection will be treated later.

Velocity Correlation Function

Perhaps the simplest example of a time correlation function is the velocity correlation function of a single particle in a fluid, $\langle v(t)v(t')\rangle$, where $v(t)$ is the velocity of that particle at time t. One reason for interest in this time correlation function is its connection with the self-diffusion coefficient D. There are many ways to show this connection. A particularly easy one starts with the one-dimensional diffusion equation for the space (x) and time (t) dependence of the concentration $C(x, t)$ of a tagged particle,

$$\frac{\partial}{\partial t} C(x,t) = D \frac{\partial^2}{\partial x^2} C(x,t). \qquad (1.17)$$

Suppose that the tagged particle starts out initially at $x = 0$. Then the concentration will change from an initial delta function in x to a spread-out Gaussian function of x. By symmetry, the mean displacement is zero. The mean squared displacement at time t can be found by multiplying the diffusion equation by x^2 and integrating over x,

$$\frac{\partial}{\partial t}\langle x^2 \rangle = \int dx x^2 \frac{\partial}{\partial t} C(x,t)$$

$$= D \int dx x^2 \frac{\partial^2}{\partial x^2} C(x,t)$$

$$= 2D \int dx C(x,t) = 2D. \qquad (1.18)$$

The last line comes from integrating by parts and by recognizing that the concentration is normalized to unity. On integrating over time, this result leads to the well-known Einstein formula for diffusion in one dimension, $\langle x^2 \rangle = 2Dt$.

Now we make a statistical mechanical theory of the same quantity. The net displacement of the particle's position during the interval from 0 to t is

$$x(t) = \int_0^t ds\, v(s), \qquad (1.19)$$

where $v(s)$ is the velocity of the particle at time s. The ensemble average of the mean squared displacement is

$$\langle x^2 \rangle = \left\langle \int_0^t ds_1 v(s_1) \int_0^t ds_2 v(s_2) \right\rangle = \int_0^t ds_1 \int_0^t ds_2 \langle v(s_1) v(s_2) \rangle. \qquad (1.20)$$

Note that the integral contains the correlation function of the velocity at times s_1 and s_2. Next, take the time derivative and combine two equivalent terms on the right-hand side,

$$\frac{\partial}{\partial t}\langle x^2 \rangle = 2\int_0^t ds\, \langle v(t)v(s) \rangle. \qquad (1.21)$$

The velocity correlation function is an equilibrium average and cannot depend on any arbitrary origin of the time axis. It can depend only on the time difference $t - s = u$, so that

$$\frac{\partial}{\partial t}\langle x^2 \rangle = 2\int_0^t ds\, \langle v(t-s)v(0) \rangle = 2\int_0^t du\, \langle v(u)v(0) \rangle. \qquad (1.22)$$

The velocity correlation function generally decays to zero in a short time; in simple liquids, this may be of the order of picoseconds. The diffusion equation is expected to be valid only at times much longer than a molecular time. In the limit of large t, the left-hand side approaches $2D$, and the right-hand side approaches a time integral from zero to infinity, so we have derived the simplest example of the relation of a transport coefficient to a time correlation function,

$$D = \int_0^\infty dt\, \langle v(t)v(0) \rangle. \qquad (1.23)$$

10 NONEQUILIBRIUM STATISTICAL MECHANICS

The three-dimensional version can be obtained by summing over x, y, and z displacements and is

$$D = \frac{1}{3}\int_0^\infty dt \langle \mathbf{V}(t) \cdot \mathbf{V}(0) \rangle, \tag{1.24}$$

where $\mathbf{V}$ is the vector velocity.

1.3 Correlation Functions and Brownian Motion

The Langevin equation and the fluctuation-dissipation theorem can be used to find expressions for various time correlation functions.

Velocity Correlation Function

The first example is to obtain the velocity correlation function of a Brownian particle. In this example, it is instructive to calculate both the equilibrium ensemble average and the long-time average.

Calculating the equilibrium ensemble average involves both an average over noise and an average over the initial velocity. The noise average leads to

$$\langle v(t) \rangle_{\text{noise}} = e^{-(\zeta/m)t}v(0). \tag{1.25}$$

Now we multiply by $v(0)$ and average over initial velocity,

$$\langle v(t)v(0) \rangle_{\text{eq}} = \frac{kT}{m}e^{-(\zeta/m)t}. \tag{1.26}$$

This holds only for $t > 0$ because the Langevin equation is valid only for positive times.

We expect that the velocity correlation function is actually a function of the absolute value of t, but to see this from the Langevin equation we have to go to the long time average. This calculation starts with a record of the time dependence of the velocity $v(t)$ over a very long time interval τ. Then the velocity correlation function can be obtained from the long time average,

$$\langle v(t)v(t') \rangle_{\text{time}} = \frac{1}{\tau}\int_0^\tau ds\, v(t+s)v(t'+s). \tag{1.27}$$

The instantaneous velocity at time t is determined by its initial value and by an integral over the noise. We assume that the initial time is the infinite past, so that the contribution from the initial value of the velocity has decayed to zero, and the instantaneous velocity is determined only by the noise. Then with a slight rearrangement of the time integral, we obtain

$$v(t) = \int_0^\infty e^{-(\zeta/m)u} \delta F(t-u) \, du/m. \qquad (1.28)$$

Now the velocity correlation function is the triple integral,

$$\langle v(t)v(t') \rangle_{\text{time}}$$
$$= \int_0^\infty du_1 \int_0^\infty du_2 \, e^{-(\zeta/m)(u_1+u_2)} \frac{1}{\tau} \int_0^\tau ds \, \frac{1}{m^2} \delta F(t-u_1+s) \delta F(t-u_2+s)$$
$$= \int_0^\infty du_1 \int_0^\infty du_2 \, e^{-(\zeta/m)(u_1+u_2)} \frac{1}{\tau} \int_0^\tau ds \, \frac{1}{m^2} 2B\delta(t-u_1-t'+u_2). \qquad (1.29)$$

The product of two random force factors has been replaced by its average. The integral over s can be done immediately. The delta function removes another integral, and the last one can be done explicitly, leading to

$$\langle v(t)v(t') \rangle = \frac{m}{2\zeta} e^{-\zeta|t-t'|/m} \frac{2B}{m^2}. \qquad (1.30)$$

Note that when the time correlation function is calculated this way, the absolute value of the time difference comes in automatically. On using the fluctuation-dissipation theorem, this leads to the final expression for the velocity correlation function,

$$\langle v(t)v(t') \rangle_{\text{time}} = \frac{kT}{m} e^{-\zeta|t-t'|/m}. \qquad (1.31)$$

The time average of the product of two velocities is the same as the equilibrium ensemble average. This is what one expects of an ergodic system. One point of this derivation is to show that observation of time dependent fluctuations over a long time interval can be used to learn about friction.

Mean Squared Displacement

Another application of the general solution of the Langevin equation is to find the mean squared displacement of the Brownian particle. The actual displacement is

$$\Delta x(t) = \int_0^t dt' \, v(t'). \qquad (1.32)$$

To find $\langle \Delta x(t)^2 \rangle_{\text{eq}}$, we start with

$$v(t) = e^{-\zeta t/m}v(0) + \int_0^t dt' e^{-\zeta(t-t')/m}\delta F(t')/m \tag{1.33}$$

and then do the averages. Since the calculation is just like earlier ones, it will be left for the reader. The result is

$$\langle \Delta x(t)^2 \rangle_{eq} = 2\frac{kT}{\zeta}\left[t - \frac{m}{\zeta} + \frac{m}{\zeta}e^{-\zeta t/m}\right]. \tag{1.34}$$

At short times, the mean squared displacement increases quadratically with time. This is the inertial behavior that comes from the initial velocity. At long times, the effects of the noise are dominant, and the mean squared displacement increases linearly with time,

$$\langle \Delta x(t)^2 \rangle \to 2\frac{kT}{\zeta}t. \tag{1.35}$$

Einstein's formula for the mean squared displacement of a diffusing particle is $2Dt$ where D is the self-diffusion coefficient of the Brownian particle. Thus we obtain Einstein's expression for the self-diffusion coefficient,

$$D = \frac{kT}{\zeta}. \tag{1.36}$$

When Stokes' law is used for the friction coefficient, the result is called the Stokes-Einstein formula. This also is a prototype of may similar expressions to be encountered later.

Dipole-Dipole Correlation Function

Many time correlation functions are related to spectroscopic measurements. For example, the frequency dependence of the optical absorption coefficient of a substance is determined by the time correlation function of its electric dipole moment. The derivation of this connection, which will be presented in Section 3.2, is an exercise in applying the quantum mechanical "Golden Rule". The result of the derivation is quite simple, especially in the classical limit where $\hbar\omega/kT \ll 1$.

Then the absorption coefficient $\alpha(\omega)$ at frequency ω is

$$\alpha(\omega) = \frac{2\pi\omega^2\beta}{3nc}\int_{-\infty}^{\infty} dt\, e^{-i\omega t}\langle \mathbf{M}(0)\cdot\mathbf{M}(t)\rangle_{eq}. \tag{1.37}$$

In the coefficient, c is the velocity of light in vacuum, and n is the index of refraction. $\mathbf{M}(t)$ is the total electric dipole moment of the system at time t. The absorption coefficient is proportional to the spectral density of the dipole-dipole time correlation function.

BROWNIAN MOTION AND LANGEVIN EQUATIONS

Suppose the system being investigated is a single rigid dipolar molecule. Then **M** is just its permanent dipole moment. It has a constant magnitude μ and a time-dependent orientation specified by the unit vector $\mathbf{u}(t)$, so that

$$\langle \mathbf{M} \cdot \mathbf{M}(t) \rangle_{eq} = \mu^2 \langle \mathbf{u}(0) \cdot \mathbf{u}(t) \rangle_{eq}. \tag{1.38}$$

If the motion is constrained to the xy plane, then it is convenient to represent the orientational vector by the angle θ,

$$\mathbf{u}(t) = (\cos \theta(t), \sin \theta(t)) \to e^{i\theta(t)}, \tag{1.39}$$

and the time correlation function of the orientations $\mathbf{u}(0)$ and $\mathbf{u}(t)$ in two dimensions can be written as

$$\langle \mathbf{u}(0) \cdot \mathbf{u}(t) \rangle_{eq} \to \langle e^{-i\theta(0)} e^{i\theta(t)} \rangle_{eq}. \tag{1.40}$$

We can calculate this quantity using the Langevin equation for rotational Brownian motion. The position x is replaced by the angle θ, the velocity v by the angular velocity Ω, and the mass m by the moment of inertia I,

$$\frac{d\theta}{dt} = \Omega, \quad I\frac{d\Omega}{dt} = -\zeta\Omega + \delta F(t) \tag{1.41}$$

and

$$\langle \delta F(t) \delta F(t') \rangle = 2\zeta k T \delta(t-t'). \tag{1.42}$$

Then, as in eq. (1.34), the equilibrium mean squared change in angle as a function of time is

$$\langle \Delta\theta(t)^2 \rangle_{eq} = 2\frac{kT}{\zeta}\left[t - \frac{I}{\zeta} + \frac{I}{\zeta} e^{-(\zeta/I)t} \right]. \tag{1.43}$$

The orientational time correlation function is

$$C(t) = \langle e^{-i\theta(0)} e^{i\theta(t)} \rangle_{eq} = \langle e^{i\Delta\theta(t)} \rangle_{eq}. \tag{1.44}$$

But $\Delta\theta(\tau)$ is linear in the noise and in the initial angular velocity, and both of these have a Gaussian distribution. (This is explained further in Appendix 2, which surveys some properties of Gaussian distributions.) Then $\Delta\theta(\tau)$ has a Gaussian distribution with a zero mean value and a second moment given by eq. (1.43), and we can use the general formula for any Gaussian average,

$$\langle \exp(iax) \rangle = \exp\left(ia\bar{x} - \frac{1}{2}a^2 \langle (x-\bar{x})^2 \rangle \right). \tag{1.45}$$

Then the time correlation function is

$$C(t) = \exp-\frac{1}{2}\langle \Delta\theta(t)^2 \rangle_{eq}. \tag{1.46}$$

At long times this decays exponentially,

$$C(t) \to \exp\left(-\frac{kT}{\zeta}t \right). \tag{1.47}$$

1.4 Brownian Motion of Other Variables

The preceding discussion started with the Brownian motion of a heavy particle, but the ideas have a much wider applicability. Another example is the kinetics of a first-order isomerization reaction between two species called A and B. For convenience, we use the same symbols, A and B, for the total number of molecules of each species that are present in a unit volume of the system. In a laboratory experiment, these are macroscopic quantities, perhaps of the order of Avogadro's number. The basic rate equations are

$$\frac{dA}{dt} = -k_1 A + k_2 B$$

$$\frac{dB}{dt} = -k_2 B + k_1 A, \tag{1.48}$$

and they have the equilibrium solutions A_{eq}, B_{eq}. The sum $A + B$ is constant in time, so that we can replace the two equations with a single one. The deviation of A from equilibrium is denoted by C, and because of conservation, the deviation of B from equilibrium is $-C$,

$$A = A_{eq} + C, \quad B = B_{eq} - C. \tag{1.49}$$

We use the equilibrium condition,

$$k_1 A_{eq} = k_2 B_{eq} \tag{1.50}$$

so that the deviation C satisfies

$$\frac{dC}{dt} = -(k_1 + k_2)C. \tag{1.51}$$

A macroscopic deviation from equilibrium decays exponentially. Now we use the "regression hypothesis" of L. Onsager (1931); this asserts that small fluctuations decay on the average in exactly the same way as macroscopic deviations from equilibrium. (This is not really a hypothesis—it seems to always be true.) Then the time correlation function of the equilibrium fluctuations in particle number is

$$\langle C(t)C(t')\rangle_{\text{time}} = \langle C^2\rangle_{\text{eq}} e^{-(k_1+k_2)|t-t'|}. \tag{1.52}$$

Equation (1.51) requires that C must decay to zero at long times; but we know that if this reacting system comes to thermal equilibrium, there are still thermal fluctuations in C, and in particular the mean squared deviation (determined by statistical thermodynamics) $\langle C^2\rangle_{\text{eq}}$ is of the order of Avogadro's number and cannot vanish. This situation is exactly like what we saw in connection with the Brownian particle. To account for the fluctuations, a "random force" or noise term $\delta F(t)$ must be added to the basic kinetic equation,

$$\frac{dC}{dt} = -(k_1+k_2)C + \delta F(t), \tag{1.53}$$

and to have the correct equilibrium behavior, we must impose the condition

$$\langle \delta F(t)\delta F(t')\rangle = 2(k_1+k_2)\langle C^2\rangle_{\text{eq}} \delta(t-t'). \tag{1.54}$$

This is evidently another version of the fluctuation-dissipation theorem. Observation of particle number fluctuations over a very long time can be used to find a rate constant.

Several Variables

At this point, it should be clear than any linear dissipative equation will lead to a similar Langevin equation and a corresponding fluctuation-dissipation theorem. The general treatment is more complex because of the possibility of both dissipative and oscillatory behavior and will be handled using a vector-matrix notation. The general treatment will be followed by an illustrative example, the Brownian motion of a harmonic oscillator.

We consider a set of dynamical variables $\{a_1, a_2, \ldots\}$ denoted by the vector $\mathbf{a}$, and the Langevin equation

$$\frac{\partial a_j}{\partial t} = \sum_k \Theta_{jk} a_k + F_j(t), \tag{1.55}$$

or in matrix form,

$$\frac{\partial \mathbf{a}}{\partial t} = \Theta \cdot \mathbf{a} + \mathbf{F}(t), \quad (1.56)$$

in which Θ is a matrix and $\mathbf{F}(t)$ is a random force vector. (To save space, the extra δ will be dropped from $\delta\mathbf{F}$.) The strength of the noise is given by

$$\langle F_j(t)F_k(t')\rangle = 2B_{jk}\delta(t-t') \quad (1.57)$$

or

$$\langle \mathbf{F}(t)\mathbf{F}(t')\rangle = 2\mathbf{B}\delta(t-t'), \quad (1.58)$$

where $\mathbf{B}$ is by definition a symmetric matrix.

Θ can be diagonalized by a similarity transformation. If it has a zero eigenvalue, the corresponding eigenvector corresponds to a dynamical constant of the motion. We assume that all such quantities have been removed from the set $\mathbf{a}$. For a system that approaches equilibrium at long times, all eigenvalues of Θ must have negative real parts; however, they can be complex.

To obtain the analog of the fluctuation-dissipation theorem for this Langevin equation, we integrate, omitting the initial value term that decays to zero at long times. The result is

$$\mathbf{a}(t) = \int_0^t ds\, e^{(t-s)\Theta} \cdot \mathbf{F}(s). \quad (1.59)$$

Now we form the matrix $\langle \mathbf{a}(t)\mathbf{a}(t)\rangle$, giving proper attention to the transpose (denoted by $\dagger$),

$$\langle \mathbf{a}(t)\mathbf{a}(t)\rangle = \int_0^t ds \int_0^t ds'\, e^{(t-s)\Theta} \cdot \langle \mathbf{F}(s)\mathbf{F}(s')\rangle \cdot e^{(t-s')\Theta^\dagger}$$

$$= \int_0^t ds\, e^{(t-s)\Theta} \cdot 2\mathbf{B} \cdot e^{(t-s)\Theta^\dagger}. \quad (1.60)$$

In the limit of very large time, this second moment must approach its equilibrium value, denoted by $\mathbf{M}$,

$$\langle \mathbf{aa}\rangle_{\text{eq}} = \mathbf{M} = 2\int_0^\infty dt\, e^{t\Theta} \cdot \mathbf{B} \cdot e^{t\Theta^\dagger}. \quad (1.61)$$

To evaluate the time integral, we first construct the symmetrized quantity $\Theta \cdot \mathbf{M} + \mathbf{M} \cdot \Theta^\dagger$ and then use the integral representation of $\mathbf{M}$,

$$\Theta \cdot \mathbf{M} + \mathbf{M} \cdot \Theta^\dagger = 2\int_0^\infty dt\, \Theta \cdot e^{t\Theta} \cdot \mathbf{B} \cdot e^{t\Theta^\dagger} + 2\int_0^\infty dt\, e^{t\Theta} \cdot \mathbf{B} \cdot e^{t\Theta^\dagger} \cdot \Theta^\dagger$$

$$= 2\int_0^\infty dt\, \frac{d}{dt} e^{t\Theta} \cdot \mathbf{B} \cdot e^{t\Theta^\dagger}$$

$$= \left(2 e^{t\Theta} \cdot \mathbf{B} \cdot e^{t\Theta^\dagger}\right)_{t=\infty} - 2\mathbf{B}. \quad (1.62)$$

The upper limit, at infinite time, vanishes because the eigenvalues of Θ all have negative real parts. So we have derived the fluctuation-dissipation theorem:

$$\Theta \cdot \mathbf{M} + \mathbf{M} \cdot \Theta^\dagger = -2\mathbf{B}. \tag{1.63}$$

Note that by their definition as second moments, $\mathbf{B}$ and $\mathbf{M}$ are symmetric, but Θ is not. According to the last equation, the product $\Theta \cdot \mathbf{M}$ has a symmetric part that is related to $\mathbf{B}$. But it can also have an antisymmetric part that has no relation to $\mathbf{B}$. It has become conventional to write Θ in the form

$$\Theta = i\Omega - \mathbf{K}. \tag{1.64}$$

The fluctuation-dissipation theorem requires both a symmetry, involving $\mathbf{K}$, and an antisymmetry, involving Ω,

$$\mathbf{B} = \mathbf{K} \cdot \mathbf{M} = \mathbf{M} \cdot \mathbf{K}^\dagger, \tag{1.65}$$

and

$$i\Omega \cdot \mathbf{M} = -\mathbf{M} \cdot i\Omega^\dagger. \tag{1.66}$$

The reason for including the factor i in $i\Omega$ is that Ω itself typically represents a frequency, so that $i\Omega$ describes oscillatory motion. The quantity $\mathbf{K} \cdot \mathbf{M}$ is real and symmetric and describes decaying motion. The symmetry of $\mathbf{K} \cdot \mathbf{M}$ is a statement of the "reciprocal relations" found by L. Onsager (1931).

A good illustration of the many-variable Langevin equation is the Brownian motion of a harmonic oscillator. We extend the earlier treatment of Langevin equations by adding an elastic force to the frictional force. The position and momentum of the oscillator are x and p, and the explicit equations of motion are

$$\frac{dx}{dt} = \frac{p}{m}$$
$$\frac{dp}{dt} = -m\omega^2 x - \zeta \frac{p}{m} + F_p(t). \tag{1.67}$$

The noise in the momentum equation is labeled by a subscript p. Then the various vectors and matrices are

$$\mathbf{a} = \begin{pmatrix} x \\ p \end{pmatrix}, \quad \mathbf{F} = \begin{pmatrix} 0 \\ F_p(t) \end{pmatrix} \tag{1.68}$$

$$\mathbf{M} = \begin{pmatrix} \langle x^2 \rangle & 0 \\ 0 & \langle p^2 \rangle \end{pmatrix} = \begin{pmatrix} kT/m\omega^2 & 0 \\ 0 & mkT \end{pmatrix} \tag{1.69}$$

18   NONEQUILIBRIUM STATISTICAL MECHANICS

$$i\Omega = \begin{pmatrix} 0 & 1/m \\ -m\omega^2 & 0 \end{pmatrix}, \quad \mathbf{K} = \begin{pmatrix} 0 & 0 \\ 0 & \zeta/m \end{pmatrix}. \tag{1.70}$$

On multiplying out the various matrices, it is easy to see that all the consequences of the fluctuation-dissipation theorem are met.

1.5 Generalizations of Langevin Equations

Nonlinear Langevin Equations

Up to now we have discussed only linear Langevin equations. They have the great practical advantage that finding analytic solutions is easy. For example, this is how the fluctuation-dissipation theorem was derived. But one often encounters nonlinear Langevin equations in modeling physical problems. A typical example is Brownian motion of a molecular dipole in a periodic potential $U(x) = u\cos 2x$. It is customary, when constructing nonlinear Langevin equations, to assume that the friction is still linear in the velocity, and that the noise is related to the friction by the same fluctuation-dissipation theorem as in the linear case. Then the equations of motion are

$$\frac{dx}{dt} = \frac{p}{m}$$

$$\frac{dp}{dt} = -U'(x) - \zeta \frac{p}{m} + \delta F(t), \tag{1.71}$$

where the force is $F(x) = -U'(x)$, and we have restored the δ in the noise term. An explicit derivation of these equations, starting with a Hamiltonian describing interaction of a system with a harmonic oscillator heat bath, is presented in the following section.

In the linear case, the first moments $\langle x \rangle$ and $\langle p \rangle$ obey exactly the same equations as the unaveraged variables, except that the noise term is absent. But if the force $F(x)$ is not linear in x, this is no longer true and the problem is much more difficult. The average equation of motion for the average momentum $\langle p \rangle$ is

$$\frac{d\langle p \rangle}{dt} = \langle F(x) \rangle - \zeta \frac{\langle p \rangle}{m} \tag{1.72}$$

and contains the average of the force. It is generally not safe to replace the average of a nonlinear function by the same function of the average,

$$\langle F(x) \rangle \neq F(\langle x \rangle). \tag{1.73}$$

This would require, for example, that the mean squared fluctuation of x must be negligible, and that is not necessarily so. A solution of the nonlinear Langevin equation will generally involve *all* moments of x and p, $\langle x^m p^n \rangle$, and these will all be coupled together.

While nonlinear Langevin equations have a pleasant pictorial character and are amenable to easy computer simulation (where the noise is modeled using random number generators), they are very hard to treat analytically. The most practical approach is to convert the Langevin equation into a Fokker-Planck equation. This will be discussed in chapter 2.

Markovian and Non-Markovian Langevin Equations

The Langevin equations considered up to now are called "Markovian." This word, familiar in the theory of probability, has a somewhat different usage in nonequlibrium statistical mechanics. It is used here to indicate that the friction at time t is proportional to the velocity at the same time, and that the noise is delta-function correlated or "white." ("White" means that the Fourier transform of the correlation function of the noise, or its spectral density, is independent of frequency.) Real problems are often not Markovian. The friction at time t can depend on the history of the velocity $v(s)$ for times s that are earlier than t. That is, the friction may have a "memory." The friction coefficient ζ is replaced by a memory function $K(t)$, sometimes called an aftereffect function, so that the frictional force at time t becomes

$$-\zeta v(t) \to -\int_{-\infty}^{t} ds\, K(t-s) v(s), \tag{1.74}$$

or, on changing variables from s to $t-s$,

$$-\zeta v(t) \to -\int_{0}^{\infty} ds\, K(s) v(t-s). \tag{1.75}$$

If a system of this sort approaches equilibrium at long times, the fluctuation-dissipation theorem must be modified; the noise is no longer white. Problems of this kind are called non-Markovian.

A simple illustration of how non-Markovian behavior can arise is by elimination of the momentum in the Brownian motion of a harmonic oscillator. The starting equations are Markovian,

$$\frac{dx}{dt} = \frac{p}{m}$$
$$\frac{dp}{dt} = -m\omega^2 x - \zeta \frac{p}{m} + F_p(t). \tag{1.76}$$

Let us suppose that the momentum vanishes in the infinite past, $p(-\infty) = 0$. We solve the second equation for $p(t)$ by integrating from $-\infty$ to t,

$$p(t) = \int_{-\infty}^{t} ds\, e^{-\zeta(t-s)/m}(-m\omega^2 x(s) + F_p(s))$$
$$= \int_{-\infty}^{t} ds\, e^{-\zeta s/m}(-m\omega^2 x(t-s) + F_p(t-s)). \tag{1.77}$$

When this is put back into the equation for dx/dt, we obtain

$$\frac{dx(t)}{dt} = -\int_0^{\infty} ds\, K(s) x(t-s) + F_x(t), \tag{1.78}$$

where the memory function $K(s)$ and the new fluctuating force $F_x(t)$ (with a subscript "x" to distinguish it from the old $F_p(t)$) are given by

$$K(t) = \omega^2 e^{-\zeta|t|/m}, \tag{1.79}$$

$$F_x(t) = \frac{1}{m} \int_0^{\infty} ds\, e^{-\zeta s/m} F_p(t-s). \tag{1.80}$$

At equilibrium, the second moment of x is

$$\langle x^2 \rangle_{eq} = \frac{kT}{m\omega^2}. \tag{1.81}$$

Then the second moment of the new random force can be worked out explicitly, using the second moment of the old force. (It is important to remember that t' can be either smaller or larger than t.) The result of this somewhat tedious calculation is

$$\langle F_x(t) F_x(t') \rangle = \langle x^2 \rangle_{eq} K(|t-t'|). \tag{1.82}$$

This is a non-Markovian version of the fluctuation-dissipation theorem. The correlation function of the new noise is proportional to the memory function for the new friction.

In the limit of very large friction, and if we are concerned only with times much longer than m/ζ, then the memory function $K(s)$ can be approximated by a delta function having the same area,

$$K(s) \cong 2\frac{m\omega^2}{\zeta} \delta(s), \tag{1.83}$$

corresponding to Markovian friction. Then eq. (1.78) becomes an approximately Markovian Langevin equation for the position $x(t)$.

Whenever variables are eliminated from a Markovian system of equations, the result is a non-Markovian system. The converse is useful to keep in mind: If the memory decays exponentially in time, a non-Markovian system can be changed into a Markovian system by adding another variable. In the present example, adding a momentum converts eq. (1.78) into the two-variable Markovian eq. (1.76).

In this treatment of non-Markovian Brownian motion, the "history" began at $t = -\infty$, and the equations reflected that. It often happens, however, that the history begins at some specified time $t = 0$. This could be, for example, because the system has been prepared in some state at that time. Then the standard form of linear non-Markovian equations is very much like those already discussed,

$$\frac{d\mathbf{a}(t)}{dt} = i\Omega \cdot \mathbf{a}(t) - \int_0^t ds \mathbf{K}(s) \cdot \mathbf{a}(t-s) + \mathbf{F}(t), \tag{1.84}$$

and the corresponding fluctuation-dissipation theorem is, in matrix form,

$$\langle \mathbf{F}(t)\mathbf{F}(t') \rangle = \mathbf{K}(t-t') \cdot \langle \mathbf{aa} \rangle_{eq}. \tag{1.85}$$

1.6 Brownian Motion in a Harmonic Oscillator Heat Bath

It is always instructive to look at simple examples, where everything can be worked out in detail. Here is a derivation of the Langevin equation for the Brownian motion of an arbitrary nonlinear system interacting bilinearly with a harmonic oscillator heat bath. This is a prototype for many statistical mechanical models, both in classical mechanics and in quantum mechanics. It will appear several times in later sections.

The main results are an exact Langevin equation, and an explanation of the way in which averages of the random force are handled. Also we can see how Markovian behavior is an approximation to true non-Markovian behavior.

The system is described by a coordinate x and its conjugate momentum p. The heat bath is described by a set of coordinates $\{q_j\}$ and their conjugate momenta $\{p_j\}$. For simplicity, all oscillator masses are set equal to 1. The system Hamiltonian H_s is

$$H_s = \frac{p^2}{2m} + U(x), \tag{1.86}$$

and the heat bath Hamiltonian H_B includes harmonic oscillator Hamiltonians for each oscillator and a very special coupling to the system,

$$H_B = \sum_j \left(\frac{p_j^2}{2} + \frac{1}{2}\omega_j^2 \left(q_j - \frac{\gamma_j}{\omega_j^2} x \right)^2 \right), \tag{1.87}$$

in which ω_j is the frequency of the jth oscillator and γ_j measures the strength of coupling of the system to the jth oscillator. H_B consists of three parts: The first is just the ordinary harmonic oscillator Hamiltonian, specified by its frequencies; the second contains a bilinear coupling to the system, $(\Sigma_j \gamma_j q_j)x$, specified by the coupling constants; and the third contains only x and could be regarded as part of the arbitrary $U(x)$. The bilinear coupling is what makes the derivation manageable.

The equations of motion for the combined Hamiltonian $H_S + H_B$ are simple:

$$\frac{dx}{dt} = \frac{p}{m}, \qquad \frac{dp}{dt} = -U'(x) + \sum_j \gamma_j \left(q_j - \frac{\gamma_j}{\omega_j^2} x \right)$$

$$\frac{dq_j}{dt} = p_j, \qquad \frac{dp_j}{dt} = -\omega_j^2 q_j + \gamma_j x. \qquad (1.88)$$

Suppose that the time dependence of the system coordinate $x(t)$ is known. Then it is easy to solve for the motion of the heat bath oscillators, in terms of their initial values and the influence of $x(t)$,

$$q_j(t) = q_j(0) \cos \omega_j t + p_j(0) \frac{\sin \omega_j t}{\omega_j} + \gamma_j \int_0^t ds\, x(s) \frac{\sin \omega_j(t-s)}{\omega_j}. \qquad (1.89)$$

Integration by parts leads to a more useful form:

$$q_j(t) - \frac{\gamma_j}{\omega_j^2} x(t) = \left(q_j(0) - \frac{\gamma_j}{\omega_j^2} x(0) \right) \cos \omega_j t + p_j(0) \frac{\sin \omega_j t}{\omega_j}$$

$$- \gamma_j \int_0^t ds\, \frac{p(s)}{m} \frac{\cos \omega_j(t-s)}{\omega_j^2}. \qquad (1.90)$$

When this is put back into the equation for dp/dt, we obtain the formal Langevin equation

$$\frac{dp(t)}{dt} = -U'(x(t)) - \int_0^t ds\, K(s) \frac{p(t-s)}{m} + F_p(t), \qquad (1.91)$$

in which the memory function $K(t)$ is explicitly

$$K(t) = \sum_j \frac{\gamma_j^2}{\omega_j^2} \cos \omega_j t, \qquad (1.92)$$

and the "noise" $F_p(t)$ is given explicitly by

$$F_p(t) = \sum_j \gamma_j p_j(0) \frac{\sin \omega_j t}{\omega_j} + \sum_j \gamma_j \left(q_j(0) - \frac{\gamma_j}{\omega_j^2} x(0) \right) \cos \omega_j t. \qquad (1.93)$$

By carefully choosing the spectrum $\{\omega_j\}$ and coupling constants $\{\gamma_j\}$, the memory function can be given any assigned form. For example, if

BROWNIAN MOTION AND LANGEVIN EQUATIONS

the spectrum is continuous, and the sum over j is replaced by an integral, $\int d\omega\, g(\omega)$, where $g(\omega)$ is a density of states, and if γ is a function of ω, then the memory function $K(t)$ becomes a Fourier integral,

$$K(t) = \int_0^\infty d\omega g(\omega) \frac{\gamma(\omega)^2}{\omega^2} \cos\omega t. \tag{1.94}$$

Further, if $g(\omega)$ is proportional to ω^2 and γ is a constant, then $K(t)$ is proportional to $\delta(t)$ and the resulting Langevin equation is Markovian.

The "noise" $F_p(t)$ is defined in terms of the initial positions and momenta of the bath oscillators and is therefore in principle a known function of time. However, if the bath has a large number of independent degrees of freedom, then the noise is a sum containing a large number of independent terms, and because of the central limit theorem, we can expect that its statistical properties are simple.

Suppose, for example, that a large number of computer simulations of this system are done. In each simulation, the bath initial conditions are taken from a distribution,

$$f_{\text{eq}}(p,q) \propto \exp(-H_B/kT), \tag{1.95}$$

in which the bath is in thermal equilibrium with respect to a frozen or constrained system coordinate $x(0)$. Then the averages of q and p are

$$\left\langle q_j(0) - \frac{\gamma_j}{\omega_j^2} x(0) \right\rangle = 0, \quad \langle p_j(0) \rangle = 0. \tag{1.96}$$

Since the noise is a linear combination of these quantities, its average value is zero. The second moments are

$$\left\langle \left(q_j(0) - \frac{\gamma_j}{\omega_j^2} x(0) \right)^2 \right\rangle = \frac{kT}{\omega_j^2}, \quad \langle p_j(0)^2 \rangle = kT. \tag{1.97}$$

There are no correlations between the initial values for different js. Then by direct calculation, using trigonometric identities, one sees immediately that there is a fluctuation-dissipation theorem,

$$\langle F_p(t) F_p(t') \rangle = kT K(t-t'). \tag{1.98}$$

Because the noise is a linear combination of quantities that have a Gaussian distribution, the noise is itself a Gaussian random variable. If the heat bath has been constructed so that the memory function is a delta function, then the noise is white or Markovian. This model justifies all the assumptions that were made about Langevin equations earlier.

24 NONEQUILIBRIUM STATISTICAL MECHANICS

In this example, the fluctuation-dissipation theorem was obtained for a rather specific kind of initial distribution of states. It may not work out so simply for a different initial distribution. One must remember that the distinction between what we call "systematic behavior" and what we call "noise" can be arbitrary; it depends on how we decide to define averages. Noise is not an intrinsic property of a material; it is determined by the experiment used to measure it.

1.7 Heavy Mass in a Harmonic Lattice

Another very instructive model of Brownian motion is due to R. J. Rubin (1960). The model is a one-dimensional harmonic lattice in which one particle is much heavier than the rest. The heavy particle appears to behave like a freely moving Brownian particle with a frictional force proportional to its velocity. This section presents a calculation of the heavy particle's velocity correlation function.

All particles except one have the same mass m. The exceptional particle has mass M. The coordinate and velocity of the jth particle are x_j and v_j, where j goes from 0 to $N - 1$. Periodic boundary conditions are used, so that $x_N = x_0$. Later the limit of very large N will be taken. Nearest neighbor particles are connected by harmonic springs so that the energy is

$$E = \frac{M}{2} v_0^2 + \sum_{j=1}^{N-1} \frac{m}{2} v_j^2 + \sum_{j=0}^{N-1} \frac{K}{2}(x_j - x_{j+1})^2. \tag{1.99}$$

The equations of motion are

$$[m + (M - m)\delta_{j0}]\ddot{x}_j = K(x_{j+1} - 2x_j + x_{j-1}). \tag{1.100}$$

The velocity correlation function (VCF), normalized to unity at $t = 0$, is the equilibrium average

$$C(t) = \frac{\langle v_0(t)v_0(0) \rangle}{\langle v_0^2 \rangle}. \tag{1.101}$$

Because the equations of motion are linear, the position and velocity at time t are linear combinations of initial positions and velocities. The equilibrium average of the product of a coordinate and a velocity vanishes, and the velocities of different particles are not correlated,

$$\langle x_j v_0 \rangle = 0, \qquad \langle v_j v_0 \rangle = \frac{k_B T}{M} \delta_{j0}. \tag{1.102}$$

As long as we want only the velocity correlation function, we do not have to solve the equations of motion for an arbitrary initial condition;

BROWNIAN MOTION AND LANGEVIN EQUATIONS 25

it is enough to set all initial coordinates and velocities except $v_0(0)$ equal to zero; their contributions will vanish anyway.

The equations of motion are conveniently solved by taking Laplace transforms (appendix 3). The Laplace transform of the jth coordinate is

$$\hat{x}_j(z) = \int_0^\infty dt\, e^{-zt} x_j(t), \tag{1.103}$$

and the equations of motion for this special choice of initial condition are

$$[m + (M-m)\delta_{j0}](z^2 \hat{x}_j - \delta_{j0} v_0(0)) = K(\hat{x}_{j+1} - 2\hat{x}_j + \hat{x}_{j-1}). \tag{1.104}$$

The structure of the potential energy suggests a normal mode transformation to new coordinates q_k,

$$x_k = \frac{1}{\sqrt{N}} \sum_{j=0}^{N-1} q_j \exp\left(\frac{2\pi i}{N} jk\right), \tag{1.105}$$

which has the inverse transformation

$$q_k = \frac{1}{\sqrt{N}} \sum_{j=0}^{N-1} x_j \exp\left(-\frac{2\pi i}{N} jk\right). \tag{1.106}$$

After some rearrangement, the transformed equations are

$$\hat{q}_k = \frac{1}{\sqrt{N}} \frac{1}{z^2 + \omega_k^2} \left[\frac{M}{m} v_0(0) - \frac{M-m}{m} z^2 \hat{x}_0\right], \tag{1.107}$$

where the normal mode frequencies are

$$\omega_k^2 = \frac{2k}{m}\left(1 - \cos\frac{2\pi k}{N}\right). \tag{1.108}$$

A further summation over k leads to

$$\hat{x}_0 = \frac{1}{N} \sum_k \frac{1}{z^2 + \omega_k^2} \left[\frac{M}{m} v_0(0) - \frac{M-m}{m} z^2 \hat{x}_0\right]. \tag{1.109}$$

The sum will be denoted by

$$\hat{\phi}(z) = \frac{1}{N} \sum_0^{N-1} \frac{1}{z^2 + \omega_k^2}. \tag{1.110}$$

This quantity is particularly simple in the limit of large N. Change variables from k to θ, and replace the sum over k by an integral over θ,

$$\hat{\phi}(z) \cong \frac{1}{2\pi}\int_0^{2\pi}d\theta \frac{1}{z^2+(2K/m)(1-\cos\theta)}$$
$$= \frac{1}{z}\frac{1}{\sqrt{z^2+4K/m}}. \qquad (1.111)$$

When eq. (1.109) is solved for $\hat{x}_0$, one gets

$$\hat{x}_0 = \frac{(1+Q)\hat{\phi}}{1+Qz^2\hat{\phi}}v_0(0), \qquad (1.112)$$

where Q is defined as

$$Q = \frac{M-m}{m}. \qquad (1.113)$$

The transform of the velocity is $z\hat{x}_0$, so the transform of the normalized velocity correlation function is

$$\hat{C}(z) = \frac{(1+Q)}{1+Qz^2\hat{\phi}(z)}z\hat{\phi}(z). \qquad (1.114)$$

In the large N limit, the approximate expression for $\hat{\phi}$ found earlier leads to an algebraic function of z,

$$\hat{C}(z) = \frac{1+Q}{Qz+\sqrt{z^2+4K/m}}. \qquad (1.115)$$

The short time behavior of $C(t)$ can be found from the large z expansion,

$$\hat{C}(z) = \frac{1}{z} - \frac{2K}{Mz^3}+\cdots. \qquad (1.116)$$

(Note the change from m to M; at short times, the inertial motion of the heavy mass dominates.) Then at short times, C is

$$C(t) = 1 - \frac{K}{M}t^2+\cdots. \qquad (1.117)$$

The quadratic dependence on t is a natural consequence of time-reversal symmetry.

The inverse Laplace transform can be found in tables if $Q = 0$ or 1. Otherwise, there are no known inverse transforms in terms of standard functions. When $Q = 0$, one finds the Bessel function of order 0,

$$C(t) = J_0(2t\sqrt{K/m}) \quad (M = m). \tag{1.118}$$

When $Q = 1$, the result is another Bessel function, of order 1,

$$C(t) = \frac{J_1(2t\sqrt{K/m})}{t\sqrt{K/m}} \quad (M = 2m). \tag{1.119}$$

After some algebraic rearrangement, the transform of the velocity correlation function may be written in the memory function form,

$$\hat{C}(z) = \frac{1}{z + \frac{m}{M}\hat{k}(z)}, \quad \hat{k}(z) = \sqrt{z^2 + 4K/m} - z. \tag{1.120}$$

On inverting the transforms, this is equivalent to

$$\frac{d}{dt}C(t) = -\frac{m}{M}\int_0^t ds\, k(s) C(t-s), \tag{1.121}$$

and the time dependent $k(s)$ (which does not depend on M) is

$$k(s) = \frac{4K}{m}\frac{J_1(2s\sqrt{K/m})}{s}. \tag{1.122}$$

Equations similar to eq. (1.121) occur frequently in nonequilibrium statistical mechanics. There is a convolution, with a memory function that has a short life time, in this case of the order of $(m/K)^{1/2}$. The convolution integral has a coefficient that can be very small, in this case of the order of m/M. In the heavy mass limit, the time derivative of $C(t)$ is small, and this suggests a Markovian approximation to eq. (1.121), where $C(t-s)$ is replaced by $C(t)$ and the integral is extended to infinity,

$$\frac{d}{dt}C(t) \approx -\frac{m}{M}\int_0^\infty ds\, k(s) C(t). \tag{1.123}$$

The infinite time integral is

$$\int_0^\infty ds\, k(s) = \sqrt{4K/m} = \frac{1}{t_0}. \tag{1.124}$$

Then the velocity correlation function decays exponentially on a time scale of the order of $M/m\, t_0$,

$$C(t) \approx \exp\left(-\frac{m}{M}\frac{t}{t_0}\right). \tag{1.125}$$

28 NONEQUILIBRIUM STATISTICAL MECHANICS

In problems of this sort, with a short memory and a small coefficient, one generally finds approximate exponential decay. But one would like to know what the limitations are on the approximation.

One can always invert a Laplace transform by means of a contour integral in the complex plane,

$$C(t) = \frac{1}{2\pi i} \int_{\varepsilon - i\infty}^{\varepsilon + i\infty} e^{tz} \frac{1+Q}{Qz + \sqrt{z^2 + 4K/m}}. \qquad (1.126)$$

Evaluating contour integrals is an exercise in complex variable theory. Rubin has done the complete calculation; we will not repeat it here. He found that there is a small correction to the exponential decay at very long times, which is bounded by

$$\left| C(t) - \exp\left(-\frac{m}{M} \frac{t}{t_0} \right) \right| \le \frac{m}{M} \sqrt{2} \qquad (1.127)$$

and which decays more slowly than exponentially. The exponential decay dominates as long as

$$t \ll \frac{M}{m} \ln \frac{M}{m} t_0. \qquad (1.128)$$

If $M = 10^4 m$, the correction to the exponential has the same order of magnitude as the exponential after about nine relaxation times.

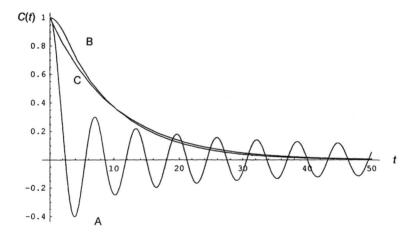

Figure 1.7.1 The velocity correlation function $C(t)$ as a function of time t. Curve A is the Bessel function for equal masses. Curve B is the result of numerical inversion of the Laplace transform when $M = 10m$. Curve C is the exponential $\exp(-t/10)$.

Sometimes one can invert a Laplace transform numerically (see Appendix 3). Figure 1.7.1 shows $C(t)$ for the equal mass case $M = m$, where A is $J_0(t)$, B is the numerical inversion for $M = 10m$, and C is the exponential $\exp(-t/10)$. Note that the asymptotic exponential works quite well except for small t.

The preceding analysis was based on the limit of large N. Two kinds of correction must be made if N is finite. One is that the magnitude of the velocity correlation function is changed by terms of the order of $1/N$. The other correction limits the time over which the large N limit applies. When the particle at $j = 0$ moves, it sends out sound waves that carry away energy and information. With periodic boundary conditions, these sound waves will eventually come back to influence that particle; in mathematical language, the motion is "almost periodic" and has recurrences. If N is large, recurrences occur only at times very much longer than the exponential relaxation time. So it is quite reasonable to take advantage of the large N limit.

2

Fokker-Planck Equations

2.1 Liouville Equation in Classical Mechanics

The foundations of nonequilibrium statistical mechanics are based on the Liouville equation. Many of the common methods for handling practical problems in nonequilibrium statistical mechanics, methods that will be described in the next few sections, either avoid the Liouville equation entirely or replace it by approximations. This is generally a reasonable thing to do; simple and approximate methods have an enormous value in science. Ultimately, however, the approximate methods must be connected with more exact and rigorous ones. This section presents a derivation of the Liouville equation in classical mechanics and shows how symbolic operator solutions of the Liouville equation can be used to deal with the properties of time correlation functions. The Liouville equation is associated with Hamiltonian dynamics; the corresponding equation associated with Langevin dynamics, called the Fokker-Planck equation, is discussed in the next section.

In classical mechanics, a system is fully specified by its coordinates and conjugate momenta. There are usually many of these; for notational convenience, the set of all coordinates will be denoted by the single symbol $\mathbf{q}$, and the conjugate momenta by the single symbol $\mathbf{p}$. The pair $(\mathbf{p}, \mathbf{q})$ gives the location of the system in its phase space, or the "phase point." Further, when there is no need to be more specific, this point is denoted for greater brevity by $\mathbf{X}$. The dynamical behavior of the system

FOKKER-PLANCK EQUATIONS 31

is determined by its Hamiltonian $H(\mathbf{p}, \mathbf{q})$ or $H(\mathbf{X})$. For now, this is taken to be independent of time.

The motion of the system in phase space is governed by Hamilton's equations,

$$\frac{\partial \mathbf{q}}{\partial t} = \frac{\partial H}{\partial \mathbf{p}}, \qquad \frac{\partial \mathbf{p}}{\partial t} = -\frac{\partial H}{\partial \mathbf{q}}. \tag{2.1}$$

In principle, this set of first-order differential equations determines the path or trajectory $\mathbf{X}_t$ in phase space that passes through a given phase point $\mathbf{X}$ at time $t = 0$. The state of the system at any time fully determines its state at all other times, both future and past. The one-dimensional harmonic oscillator provides an easy illustration. The Hamiltonian is

$$H(p, q) = \frac{1}{2m} p^2 + \frac{1}{2} m\omega^2 q^2, \tag{2.2}$$

the equations of motion are

$$\frac{dq}{dt} = \frac{p}{m}, \qquad \frac{dq}{dt} = -m\omega^2 q, \tag{2.3}$$

and their solution, valid for all t, is

$$q(t) = \cos(\omega(t - t_0)) q(t_0) + \frac{\sin(\omega(t - t_0))}{m\omega} p(t_0)$$

$$p(t) = -\omega m \sin(\omega(t - t_0)) q(t_0) + \cos(\omega(t - t_0)) p(t_0). \tag{2.4}$$

Aside from this special case, one can seldom solve Hamilton's equations exactly. Further, one learns from research on chaotic dynamical systems that solutions may be extremely sensitive to small changes in the initial state, so that "exact in principle" does not always mean "exact in practice." It may be very hard to give a precise prediction of the future behavior of a given initial state. However, because the present concern is with formalism, and not with practical calculations, this difficulty may be ignored.

The Liouville Equation

In classical statistical mechanics, averages are determined by the phase space distribution function (or phase space density) $f(\mathbf{p}, \mathbf{q}, t)$ or $f(\mathbf{X}, t)$. The probability of finding the system state in the region $d\mathbf{X}$ around the point $\mathbf{X}$ at time t is $f(\mathbf{X}, t)d\mathbf{X}$. Probability is conserved; the total probability that the system is *somewhere* is unity at all times,

NONEQUILIBRIUM STATISTICAL MECHANICS

$$\int d\mathbf{X} f(\mathbf{X}, t) = 1 \quad \text{(all } t\text{)}. \tag{2.5}$$

As in fluid mechanics or electrodynamics, whenever an integral of a quantity $\rho(\mathbf{X})$ over an entire domain is conserved, there is generally a conservation law of the form

$$\frac{\partial \rho}{\partial t} = -\frac{\partial}{\partial \mathbf{X}} \cdot (\rho \mathbf{V}), \tag{2.6}$$

where ρ is a density, $\mathbf{V}$ is a velocity, and $\rho \mathbf{V}$ is a flux. In the present instance, the time rate of change of the density f is the negative of the divergence of its flux in phase space, and the correspondences are

$$\frac{\partial}{\partial \mathbf{X}} \to \left(\frac{\partial}{\partial \mathbf{q}}, \frac{\partial}{\partial \mathbf{p}} \right), \quad \mathbf{V} \to \left(\frac{d\mathbf{q}}{dt}, \frac{d\mathbf{p}}{dt} \right),$$

and

$$\frac{\partial f}{\partial t} = -\frac{\partial}{\partial \mathbf{q}} \cdot \left(\frac{d\mathbf{q}}{dt} f \right) - \frac{\partial}{\partial \mathbf{p}} \cdot \left(\frac{d\mathbf{p}}{dt} f \right), \tag{2.7}$$

On using Hamilton's equations for the time derivatives of $\mathbf{p}$ and $\mathbf{q}$ and canceling out cross terms, this becomes the *Liouville equation* for the probability distribution function,

$$\frac{\partial f}{\partial t} = -\frac{\partial H}{\partial \mathbf{p}} \cdot \frac{\partial f}{\partial \mathbf{q}} + \frac{\partial H}{\partial \mathbf{q}} \cdot \frac{\partial f}{\partial \mathbf{p}}. \tag{2.8}$$

It is often convenient to write this in an operator form. The Liouville operator is defined by

$$L = \frac{\partial H}{\partial \mathbf{p}} \cdot \frac{\partial}{\partial \mathbf{q}} - \frac{\partial H}{\partial \mathbf{q}} \cdot \frac{\partial}{\partial \mathbf{p}}, \tag{2.9}$$

so that the Liouville equation is

$$\frac{\partial}{\partial t} f = -Lf. \tag{2.10}$$

The Liouville equation has the formal operator solution

$$f(\mathbf{X}, t) = e^{-tL} f(\mathbf{X}, 0). \tag{2.11}$$

The Liouville operator is sometimes (especially in older literature) written with the imaginary factor $i = \sqrt{-1}$, so that $\partial f/\partial t = -iLf$. There seems to be no special advantage in carrying along the extra factor of i.

One property of the Liouville operator merits special comment because it is so frequently used. Consider the integral of LAf over

FOKKER-PLANCK EQUATIONS 33

the entire phase space. A multidimensional version of the basic integral,

$$\int_a^b dx \frac{d}{dx} u(x) = u(b) - u(a), \tag{2.12}$$

can be used to convert an integral over the phase space volume to an integral over the surface of that volume,

$$\int_{\text{volume}} d\mathbf{X} L A f = -\int_{\text{volume}} d\mathbf{X} \frac{d}{d\mathbf{X}} \cdot (\mathbf{V} A f) = -\oint_{\text{surface}} d\mathbf{S} \cdot (\mathbf{V} A f), \tag{2.13}$$

where $\mathbf{S}$ is a unit vector normal to the boundary. The volume integral vanishes as long as $\mathbf{V}Af$ vanishes on the boundary of phase space. Typically, the system is confined to a finite region in configuration space and has a finite energy. Then the distribution function f vanishes for coordinates outside that region, and also at very large momenta. Further, because L contains first derivatives, it can be distributed over a product, $L(Af) = (LA)f + A(Lf)$. Consequently, L is anti-self-adjoint in phase space,

$$\int d\mathbf{X} A(Lf) = -\int d\mathbf{X} (LA)f. \tag{2.14}$$

Dynamical Variables

The goal of nonequilibrium statistical mechanics is to understand the time evolution of dynamical properties of a many-body system. These may be some experimentally observable properties, such as the hydrodynamic variables (mass density, momentum density, and energy density), or they may even be some of the coordinates and momenta themselves. For the present, their exact nature is not important. What is important is that dynamical properties of a system are functions of its state.

A generic dynamical variable will be denoted by $A(\mathbf{X})$. Because the state changes with time, so does this variable; at time t, its value is $A(\mathbf{X}_t)$. Because the state at time t depends parametrically on the initial state $\mathbf{X}$, the value of the variable at time t depends on $\mathbf{X}$. This prompts some notation that may be confusing at first but is actually quite helpful. The symbol A will be used in three ways. If A does not contain any argument at all or contains explicitly only the argument $\mathbf{X}$, as in $A(\mathbf{X})$, then it denotes the variable itself. If A explicitly contains the argument t, as in $A(t)$, $A(\mathbf{X}_t)$ or $A(\mathbf{X}, t)$, then it denotes the *value* of the variable at time t as it evolved from the initial state $\mathbf{X}$. The value of $A(t)$ at $t = 0$ is A. (The same multiplicity of interpretations occurs in quantum mechanics, where, e.g., $\psi(x)$ can denote a pure quantum state as a function of position x, and $\psi(t, x)$ is a system's wave function as it evolves in time.) This notation is summarized by $A(t) = A(\mathbf{X}_t) = A(\mathbf{X}, t)$ and by $A = A(\mathbf{X}) = A(t = 0)$.

When the dynamical variable A is regarded as a function of both the time t and the initial state $\mathbf{X}$, its initial rate of change with time is a function of $\mathbf{X}$ and can be calculated from

$$\left(\frac{\partial A}{\partial t}\right)_{t=0} = \frac{\partial A}{\partial \mathbf{q}} \cdot \left(\frac{\partial \mathbf{q}}{\partial t}\right)_{t=0} + \frac{\partial A}{\partial \mathbf{p}} \cdot \left(\frac{\partial \mathbf{p}}{\partial t}\right)_{t=0}$$
$$= \left(\frac{\partial H}{\partial \mathbf{p}} \cdot \frac{\partial}{\partial \mathbf{q}} - \frac{\partial H}{\partial \mathbf{q}} \cdot \frac{\partial}{\partial \mathbf{p}}\right) A. \qquad (2.15)$$

This contains the same Liouville operator that was defined in eq. (2.9). Note that L operates on functions of the location $\mathbf{X} = (\mathbf{p}, \mathbf{q})$ in phase space. The initial rate of change is LA; the initial second time derivative is the initial rate of change of the initial first derivative, LLA, and so on. The nth initial time derivative is

$$\left(\frac{\partial^n A}{\partial t^n}\right)_{t=0} = L^n A. \qquad (2.16)$$

This can be used to generate a formal Taylor's series expansion of the time-dependent dynamical variable in powers of t,

$$A(\mathbf{X}, t) = \sum_{j=0}^{\infty} \frac{t^n}{n!} \left(\frac{\partial^n A}{\partial t^n}\right)_{t=0} = \sum_{j=0}^{\infty} \frac{t^n}{n!} L^n A(\mathbf{X}) = e^{tL} A(\mathbf{X}). \qquad (2.17)$$

This evidently is the solution of the operator equation,

$$\frac{\partial}{\partial t} A(\mathbf{X}, t) = L A(\mathbf{X}, t), \qquad A(\mathbf{X}, 0) = A(\mathbf{X}). \qquad (2.18)$$

This is the Liouville equation for the evolution of a dynamical variable. Just as the Liouville equation for the distribution function is analogous to the Schrodinger equation in quantum mechanics, this equation is analogous to the Heisenberg equation of motion.

The operator $\exp(tL)$ moves any dynamical variable along a trajectory in phase space and is sometimes called a "propagator." It has several interesting and useful properties. For example, it can be moved inside a function,

$$e^{tL} A(\mathbf{X}) = A(e^{tL} \mathbf{X}), \qquad (2.19)$$

and it can be distributed over products of functions,

$$e^{tL} A(\mathbf{X}) B(\mathbf{X}) = (e^{tL} A(\mathbf{X}))(e^{tL} B(\mathbf{X})). \qquad (2.20)$$

These identities are due to the uniqueness of the trajectory that passes through any specified phase point $\mathbf{X}$.

The phase space average of a dynamical variable A at time t is

$$\langle A, t \rangle = \int d\mathbf{X}\, A(\mathbf{X}) f(\mathbf{X}, t) = \int d\mathbf{X}\, A(\mathbf{X}) e^{-tL} f(\mathbf{X}, 0). \qquad (2.21)$$

But this is also the average of the time-dependent dynamical variable over the distribution of initial states,

$$\langle A, t \rangle = \int d\mathbf{X}\, A(\mathbf{X}, t) f(\mathbf{X}, 0) = \int d\mathbf{X}\, (e^{tL} A(\mathbf{X})) f(\mathbf{X}, 0). \qquad (2.22)$$

These two forms are equivalent because L is anti-self-adjoint in phase space. This is analogous to the Schrodinger-Heisenberg duality in quantum mechanics.

Time Correlation Functions

The Liouville operator notation provides a convenient way of manipulating equilibrium time correlation functions (TCF). Some examples are given here. The TCF of the dynamical variables A and B is

$$C_{AB}(t) = \int d\mathbf{X}\, A(\mathbf{X}, t) B(\mathbf{X}) f_{eq}(\mathbf{X}), \qquad (2.23)$$

where f_{eq} is the equilibrium distribution function. This is also

$$C_{AB}(t) = \int d\mathbf{X}\, (e^{tL} A(\mathbf{X})) B(\mathbf{X}) f_{eq}(\mathbf{X}), \qquad (2.24)$$

and by taking the adjoint, it becomes

$$C_{AB}(t) = \int d\mathbf{X}\, A(\mathbf{X}) (e^{-tL} B(\mathbf{X}) f_{eq}(\mathbf{X})). \qquad (2.25)$$

The exponential operator can be distributed over B and f, and recognizing that the equilibrium distribution function is invariant to L, we find

$$C_{AB}(t) = \int d\mathbf{X}\, A(t) B(\mathbf{X}, -t) f_{eq}(\mathbf{X}) = C_{BA}(-t). \qquad (2.26)$$

If A and B are the same quantity, their TCF is invariant to time reversal.

The time derivative of a TCF is another TCF,

$$\begin{aligned}\frac{\partial}{\partial t} C_{AB}(t) &= \int d\mathbf{X}\, \frac{\partial}{\partial t} A(\mathbf{X}, t) B(\mathbf{X}) f_{eq}(\mathbf{X}) \\ &= \int d\mathbf{X} (LA(\mathbf{X}, t)) B(\mathbf{X}) f_{eq}(\mathbf{X}) \\ &= -\int d\mathbf{X}\, A(\mathbf{X}, t)(LB(\mathbf{X})) f_{eq}(\mathbf{X}).\end{aligned} \qquad (2.27)$$

This is the TCF of A and the time derivative of B. In the same way, the second time derivative is

$$\frac{\partial^2}{\partial t^2} C_{AB}(t) = -\int d\mathbf{X} \dot{A}(\mathbf{X}, t)\dot{B}(\mathbf{X}) f_{eq}(\mathbf{X}). \qquad (2.28)$$

So, for example, the second time derivative of the velocity correlation function is the negative of the force-force correlation function,

$$\frac{\partial^2}{\partial t^2} \langle v(t)v \rangle_{eq} = -\frac{1}{m^2} \langle F(t)F \rangle_{eq}. \qquad (2.29)$$

2.2 Fokker-Planck Equations

Fokker-Planck equations are a form of Liouville equation used to treat the statistical behavior of dynamical systems with Markovian friction and Gaussian white noise. This section contains a derivation of the Fokker-Planck equation that corresponds to a given Langevin equation and some simple illustrations. Some general properties of Fokker-Planck equations will be discussed in the following section.

Earlier sections dealt with Langevin equations and their associated fluctuation-dissipation theorems. It was observed that linear Langevin equations are easy to solve and that the effects of noise are easy to work out. But it was also observed that nonlinear Langevin equations are not easy to solve; the nonlinearity, while not a problem for numerical simulations, introduces serious difficulties in analytic studies. One practical way to handle these difficulties is to construct the Fokker-Planck equation that corresponds to a given Langevin equation.

Derivation of a Fokker-Planck Equation

Let us start with a quite general Langevin equation for the dynamics of a set of variables $\{a_1, a_2, \ldots\}$ denoted for convenience by $\mathbf{a}$. At the beginning, no special requirements are imposed on the noise-free part of the dynamics, except that it is Markovian (i.e., has no memory). However, we do require that the noise is white and has a Gaussian distribution. The equations of motion are

$$\frac{da_j}{dt} = v_j(a_1, a_2, \cdots) + F_j(t), \qquad (2.30)$$

or, in abbreviated form,

$$\frac{d\mathbf{a}}{dt} = \mathbf{v}(\mathbf{a}) + \mathbf{F}(t), \qquad (2.31)$$

FOKKER-PLANCK EQUATIONS   37

where $\mathbf{v}(\mathbf{a})$ is some given function of the variables $\mathbf{a}$. The noise $\mathbf{F}(t)$ is Gaussian, with zero mean and the delta-correlated second moment matrix,

$$\langle \mathbf{F}(t)\mathbf{F}(t') \rangle = 2\mathbf{B}\delta(t-t'). \tag{2.32}$$

Rather than looking for a general solution of these equations, we ask for the probability distribution $f(\mathbf{a}, t)$ of the values of $\mathbf{a}$ at time t. Further, what we really want is the average of this probability distribution over the noise. One way to find the noise average is to start by recognizing that $f(\mathbf{a}, t)$ is a conserved quantity,

$$\int d\mathbf{a}\, f(\mathbf{a}, t) = 1, \quad \text{all } t. \tag{2.33}$$

Whenever a conservation law of this kind is encountered, we expect that the time derivative of the conserved quantity or density (in this case, $f(\mathbf{a}, t)$) is balanced by the divergence of a flux, a velocity times that density. This is the way, for example, that the Liouville equation is derived in statistical mechanics. Here the space coordinates are $\mathbf{a}$, the density at $\mathbf{a}$ is $f(\mathbf{a}, t)$, the velocity at $\mathbf{a}$ is $d\mathbf{a}/dt$, and the conservation law is

$$\frac{\partial f}{\partial t} + \frac{\partial}{\partial \mathbf{a}} \cdot \left(\frac{\partial \mathbf{a}}{\partial t} f\right) = 0. \tag{2.34}$$

On replacing the time derivative by the right-hand side of eq. (2.31), we get

$$\frac{\partial f(\mathbf{a}, t)}{\partial t} = -\frac{\partial}{\partial \mathbf{a}} \cdot (\mathbf{v}(\mathbf{a}) f(\mathbf{a}, t) + \mathbf{F}(t) f(\mathbf{a}, t)). \tag{2.35}$$

This contains a random term and is called a stochastic differential equation. We want to use it to derive an equation for the noise average of f.

The derivation is considerably simplified by using some symbolic operator manipulations. To begin, we define an operator L (analogous to the Liouville operator) by its action on any function Φ,

$$L\Phi \equiv \frac{\partial}{\partial \mathbf{a}} \cdot (\mathbf{v}(\mathbf{a})\Phi). \tag{2.36}$$

This is used to write a symbolic solution of the noise-free equation,

$$\frac{\partial f}{\partial t} = -Lf. \tag{2.37}$$

The formal or symbolic solution (as an initial value problem) is

$$f(\mathbf{a}, t) = e^{-tL} f(\mathbf{a}, 0). \tag{2.38}$$

Now we add the noise term,

$$\frac{\partial f}{\partial t} = -Lf - \frac{\partial}{\partial \mathbf{a}} \cdot \mathbf{F}(t) f. \tag{2.39}$$

One integration over time leads to the operator equation,

$$f(\mathbf{a}, t) = e^{-tL} f(\mathbf{a}, 0) - \int_0^t ds\, e^{-(t-s)L} \frac{\partial}{\partial \mathbf{a}} \cdot \mathbf{F}(s) f(\mathbf{a}, s). \tag{2.40}$$

It is important to realize that $f(\mathbf{a}, t)$ depends on the noise $\mathbf{F}(s)$ only for times s that are earlier than t. By iterating, we develop a series expansion for f in powers of the noise. Equation (2.40) is substituted back into eq. (2.39), leading to

$$\frac{\partial}{\partial t} f(\mathbf{a}, t) = -L f(\mathbf{a}, t) - \frac{\partial}{\partial \mathbf{a}} \cdot \mathbf{F}(t) f(\mathbf{a}, 0)$$
$$+ \frac{\partial}{\partial \mathbf{a}} \cdot \mathbf{F}(t) \int_0^t ds\, e^{-(t-s)L} \frac{\partial}{\partial \mathbf{a}} \cdot \mathbf{F}(s) f(\mathbf{a}, s). \tag{2.41}$$

Now we take the average over noise. The initial distribution function $f(\mathbf{a}, 0)$ is not affected by the average, so the term with a single $\mathbf{F}$ and the initial distribution function average to zero. The final term contains two explicit noise factors, $\mathbf{F}(t)$ and $\mathbf{F}(s)$, and also those noise factors that are implicit in $f(\mathbf{a}, s)$, but only with times earlier than s. The noise is Gaussian and delta-function correlated; this means that on averaging, we can pair the first factor $\mathbf{F}(t)$ with the second factor $\mathbf{F}(s)$ or with one of the implicit noise factors in $f(\mathbf{a}, s)$. (For a further explanation, see Appendix 2.) In the first case, we get $\delta(t - s)$, and is the second case we get $\delta(t - s')$ with $s' < s$. But this second case is not allowed because of the limitation to $t > s > s'$. Thus only the first two noise factors need to be paired. The average introduces a factor $\mathbf{B}$, and the delta function removes the operator $e^{-(t-s)L}$. The result is the Fokker-Planck equation for the noise-averaged distribution function $\langle f(\mathbf{a}, t) \rangle$,

$$\frac{\partial}{\partial t} \langle f(\mathbf{a}, t) \rangle = -\frac{\partial}{\partial \mathbf{a}} \cdot \mathbf{v}(\mathbf{a}) \langle f(\mathbf{a}, t) \rangle + \frac{\partial}{\partial \mathbf{a}} \cdot \mathbf{B} \cdot \frac{\partial}{\partial \mathbf{a}} \langle f(\mathbf{a}, t) \rangle. \tag{2.42}$$

The first term on the right-hand side is what one had on the absence of noise. The second term on the right-hand side accounts for the averaged effects of the noise. At this point, $\mathbf{B}$ can be any function of $\mathbf{a}$.

Several comments are in order. The derivation as given here depends explicitly on two assumptions, that the noise is Gaussian and that it is delta-function correlated. Otherwise, the factorization of the average over noise will not work. In particular, the derivation will not work for a non-Markovian Langevin equation.

Also, no fluctuation-dissipation theorem has been invoked. Nothing has been said about requiring that $\langle f(\mathbf{a}, t)\rangle$ must approach an equilibrium distribution at long times. If there is not enough friction to dampen the heating effect of the noise, we expect that the system will "run away" so that there is no long time steady state. If there is too much friction for the noise, the system will cool down and "die." In fact, not much is known in general about the long time steady state solution of an *arbitrary* Fokker-Planck equation. All that we can usually do is guess at a steady state solution, put it into the equation, and see if our guess is compatible with $\mathbf{v}(\mathbf{a})$ and $\mathbf{B}$. If a steady-state solution is found, then it implies a relation between $\mathbf{v}(\mathbf{a})$ and $\mathbf{B}$ which may be called a fluctuation-dissipation theorem.

In later uses, the angular brackets "$\langle\ \rangle$" denoting the noise average will be omitted; we will deal only with the averaged distribution.

Illustrations

The first illustration is the two-variable Brownian motion of a particle moving in the potential $U(x)$. The Langevin equations are

$$\frac{dx}{dt} = \frac{p}{m}, \qquad \frac{dp}{dt} = -U'(x) - \zeta \frac{p}{m} + F_p(t), \tag{2.43}$$

and the fluctuation-dissipation theorem is

$$\langle f_p(t) f_p(t')\rangle = 2\zeta kT \delta(t - t'). \tag{2.44}$$

The quantities that go into the general Fokker-Planck equation are

$$\mathbf{a} = \begin{pmatrix} x \\ p \end{pmatrix}, \qquad \mathbf{v}(\mathbf{a}) = \begin{pmatrix} p/m \\ -U'(x) - \zeta p/m \end{pmatrix}, \tag{2.45}$$

$$\mathbf{F}(t) = \begin{pmatrix} 0 \\ F_p(t) \end{pmatrix}, \qquad \mathbf{B} = \begin{pmatrix} 0 & 0 \\ 0 & \zeta kT \end{pmatrix}. \tag{2.46}$$

Then the Fokker-Planck equation becomes

$$\frac{\partial f}{\partial t} = -\frac{\partial}{\partial x}\frac{p}{m} f - \frac{\partial}{\partial p}(-U'(x) - \zeta p/m) f + \zeta kT \frac{\partial^2}{\partial p^2} f. \tag{2.47}$$

Note that if there is no noise or friction, the Fokker-Planck equation reduces to the standard Liouville equation for the Hamiltonian,

$$H = \frac{p^2}{2m} + U(x). \tag{2.48}$$

With noise and friction, the equilibrium solution ($\partial f/\partial t = 0$) is

$$f_{eq}(x, p) = \frac{1}{Q} e^{-H(x,p)/kT}, \quad Q = \iint dx dp \, e^{-H(x,p)/kT}, \tag{2.49}$$

where Q is the partition function at temperature T.

This Fokker-Planck equation is the starting point for many useful calculations, for example, to determine the rate at which a Brownian particle crosses a potential barrier. The corresponding equation in which the coordinate x is replaced by an angle and the momentum p by an angular momentum is useful in treating molecular reorientation in liquids.

Another example starts with the same Langevin equation, but now we assume that the relaxation time $\tau = m/\zeta$ is very much shorter than any natural time scale associated with motion in the potential $U(x)$. There are several ways to use this assumption; one was discussed earlier. Another procedure is to start with the Langevin equation,

$$m \frac{d^2 x}{dt^2} = -U'(x) - \zeta \frac{dx}{dt} + F(t). \tag{2.50}$$

We drop the second derivative on the left-hand side and rearrange to get an approximate Langevin equation for $x(t)$ alone,

$$\frac{dx}{dt} = -\frac{1}{\zeta} U'(x) + \frac{1}{\zeta} F(t). \tag{2.51}$$

This leads to a Fokker-Planck equation that is commonly called the *Smoluchowski equation*,

$$\begin{aligned}\frac{\partial f}{\partial t} &= \frac{1}{\zeta} \frac{\partial}{\partial x} U'(x) f + \frac{kT}{\zeta} \frac{\partial^2}{\partial x^2} f \\ &= D \frac{\partial}{\partial x} e^{-U(x)/kT} \frac{\partial}{\partial x} e^{U(x)/kT} f.\end{aligned} \tag{2.52}$$

This equation describes diffusion in an external potential; the diffusion coefficient is

$$D = \frac{kT}{\zeta}. \tag{2.53}$$

While the Smoluchowski equation is a correct representation of the Langevin dynamics of eq. (2.51), it is only an approximation to the Langevin equation of eq. (2.43).

2.3 About Fokker-Planck Equations

Some Properties

Fokker-Planck equations are parabolic differential equations, but of a special kind. Normally, only a few of the variables appear in the second derivative part of the equation. (The Smoluchowski equation is an important exception.) There is no guarantee of a steady state solution. These equations are generally not self-adjoint, and little is known about their mathematical properties. It is likely that they can have eigenfunctions and eigenvalues. These can be worked out in special cases, for example, a planar rigid rotor or a harmonic oscillator. But there appear to be no general theorems about the existence or completeness of eigenfunction expansions and similar questions.

The Smoluchowski equation, as an exception to the general rule, can be made self-adjoint by a trick. The substitution

$$f = \sqrt{f_{eq}}\, g \tag{2.54}$$

leads to a Schrodinger-like equation for g,

$$-\frac{\partial}{\partial t} g = D\left(-\frac{\partial^2}{\partial x^2} + U_{eff}(x)\right) g \tag{2.55}$$

$$U_{eff}(x) = \left\{\left(\frac{1}{2kT}\frac{\partial U}{\partial x}\right)^2 - \frac{1}{2kT}\frac{\partial^2 U}{\partial x^2}\right\}. \tag{2.56}$$

The original potential U has been replaced by an effective potential U_{eff} involving the force and its derivative. This transformation leads to an equation whose properties are very well known. It has real eigenvalues and eigenfunctions, which form a complete set. While finding solutions may be just as hard as in quantum mechanical problems, there are no conceptual difficulties. But for the more-general Fokker-Planck equation, the same trick does not lead to a self-adjoint equation.

As was observed in section 1.5, non-Markovian Langevin equations with exponentially decaying memory can be converted to Markovian Langevin equations by increasing the number of variables. Thus, a non-Markovian two variable Langevin equation with exponential memory gives rise to a three-variable Markovian Langevin equation and hence to a three-variable Fokker-Planck equation. By analogy with

NONEQUILIBRIUM STATISTICAL MECHANICS

non-Markovian Langevin equations, one might be tempted to write down the corresponding non-Markovian Fokker-Planck equation, that is, one with memory; this is not safe.

Averages and Adjoint Operators

Sometimes we want the full solution of a Fokker-Planck equation, but sometimes we are interested only in certain averages. These can be found by two distinct but equivalent procedures analogous to the Schrodinger-Heisenberg duality in quantum mechanics.

First, we can follow the evolution of some initial state $f(\mathbf{a}, t)$, by solving the Fokker-Planck equation,

$$\frac{\partial}{\partial t} f(\mathbf{a}, t) = \mathcal{D} f(\mathbf{a}, t) \tag{2.57}$$

where the operator $\mathcal{D}$ is given by

$$\mathcal{D} = -\frac{\partial}{\partial \mathbf{a}} \cdot \mathbf{v}(\mathbf{a}) + \frac{\partial}{\partial \mathbf{a}} \cdot \mathbf{B} \cdot \frac{\partial}{\partial \mathbf{a}}. \tag{2.58}$$

The first part of this operator is the L that appeared earlier. The second part represents the average effects of noise. The Fokker-Planck equation has a formal operator solution,

$$f(\mathbf{a}, t) = e^{\mathcal{D} t} f(\mathbf{a}, 0). \tag{2.59}$$

This can be used to get the average of any dynamical property $\phi(\mathbf{a})$ (including the special case $\phi(\mathbf{a}) = \mathbf{a}$),

$$\langle \phi, t \rangle = \int d\mathbf{a}\, \phi(\mathbf{a}) f(\mathbf{a}, t) = \int d\mathbf{a}\, \phi(\mathbf{a}) e^{\mathcal{D} t} f(\mathbf{a}, 0). \tag{2.60}$$

This may be called the "Schrodinger approach," since it focuses on the evolution of a probability distribution, and the average is taken at time t.

The second way to get the average uses the operator that is adjoint to $\mathcal{D}$ defined by

$$\int d\mathbf{a}\, \phi(\mathbf{a}) \mathcal{D} \psi(\mathbf{a}) = \int d\mathbf{a}\, \psi(\mathbf{a}) \mathcal{D}^\dagger \phi(\mathbf{a}) \tag{2.61}$$

$$\mathcal{D}^\dagger = \mathbf{v}(\mathbf{a}) \cdot \frac{\partial}{\partial \mathbf{a}} + \frac{\partial}{\partial \mathbf{a}} \cdot \mathbf{B} \cdot \frac{\partial}{\partial \mathbf{a}}. \tag{2.62}$$

Now the average can be obtained by reversing the operator in the exponent,

FOKKER-PLANCK EQUATIONS 43

$$\langle \phi, t \rangle = \int d\mathbf{a} f(\mathbf{a}, 0) e^{\mathcal{D}^\dagger t} \phi(\mathbf{a}) = \int d\mathbf{a} f(\mathbf{a}, 0) \phi(\mathbf{a}, t), \quad (2.63)$$

which contains the *defined* time-dependent variable,

$$\phi(\mathbf{a}, t) = e^{\mathcal{D}^\dagger t} \phi(\mathbf{a}). \quad (2.64)$$

This may be called the "Heisenberg approach" since it focuses on the evolution of a dynamical observable, and the average is taken over an initial distribution. The equation of motion for ϕ becomes

$$\frac{\partial}{\partial t} \phi(\mathbf{a}, t) = \left[\mathbf{v}(\mathbf{a}) \cdot \frac{\partial}{\partial \mathbf{a}} + \frac{\partial}{\partial \mathbf{a}} \cdot \mathbf{B} \cdot \frac{\partial}{\partial \mathbf{a}} \right] \phi(\mathbf{a}, t), \quad \phi(\mathbf{a}, 0) = \phi(\mathbf{a}). \quad (2.65)$$

The time dependence of $\phi(\mathbf{a}, t)$ is not what one would see in a single experiment, that is, before averaging over noise. This quantity is defined so as to give the correct time dependence after averaging over noise but before averaging over initial conditions.

Because the two operators $(\mathcal{D}, \mathcal{D}^\dagger)$ are so similar in structure, there is usually no advantage in using one instead of the other, except in formal operator manipulations. The solution of a Fokker-Planck equation and its adjoint equation are equally difficult.

Green's Function in the Linear Case

A formal solution of the Fokker-Planck equation, as an initial value problem, uses the exponential operator $\exp t\mathcal{D}$. A more-explicit solution uses the Green's function $G(\mathbf{a}, t|\mathbf{a}_0)$,

$$f(\mathbf{a}; t) = e^{t\mathcal{D}} f(\mathbf{a}; 0) = \int d\mathbf{a} G(\mathbf{a}, t|\mathbf{a}_0) f(\mathbf{a}_0, 0). \quad (2.66)$$

The Green's function satisfies the same Fokker-Planck equation, but with the special initial condition,

$$G(\mathbf{a}, t = 0|\mathbf{a}_0) = \delta(\mathbf{a} - \mathbf{a}_0). \quad (2.67)$$

When the streaming function $\mathbf{v}(\mathbf{a})$ is linear in $\mathbf{a}$,

$$\mathbf{v}(\mathbf{a}) = \Theta \cdot \mathbf{a}, \quad (2.68)$$

the Green's function can be found easily by the following procedure. First we take the Fourier transform of G (in the language of Appendix 2, we construct the moment generating function),

$$\gamma(\xi, t|\mathbf{a}_0) = \int d\mathbf{a} \, e^{i\xi \cdot \mathbf{a}} G(\mathbf{a}, t|\mathbf{a}_0). \quad (2.69)$$

Then on integrating by parts several times, the Fokker-Planck equation becomes a first-order partial differential equation,

$$\frac{\partial}{\partial t}\gamma = \xi \cdot \Theta \cdot \frac{\partial}{\partial \xi}\gamma - \xi \cdot B \cdot \xi \gamma. \tag{2.70}$$

The logarithm of γ satisfies

$$\frac{\partial}{\partial t}\ln\gamma = \xi \cdot \Theta \cdot \frac{\partial}{\partial \xi}\ln\gamma - \xi \cdot B \cdot \xi. \tag{2.71}$$

This suggests that we expand $\ln\gamma$ in powers of ξ,

$$\ln\gamma = i\alpha(t)\cdot\xi - \frac{1}{2}\xi\cdot\sigma(t)\cdot\xi. \tag{2.72}$$

The time-dependent coefficients obey simple differential equations,

$$\frac{\partial}{\partial t}\alpha = \Theta\cdot\alpha, \qquad \frac{\partial}{\partial t}\sigma = \Theta\cdot\sigma + \sigma\cdot\Theta^\dagger + 2B. \tag{2.73}$$

But the initial value of γ is simply $\exp(i\xi\cdot\mathbf{a}_0)$, so the initial values of the coefficients are

$$\alpha(0) = \mathbf{a}_0, \qquad \sigma(0) = 0. \tag{2.74}$$

These equations have solutions,

$$\alpha(t) = e^{t\Theta}\cdot\mathbf{a}_0, \qquad \sigma(t) = 2\int_0^t ds\, e^{(t-s)\Theta}\cdot B\cdot e^{(t-s)\Theta^\dagger}. \tag{2.75}$$

(The expression for σ appeared earlier, in section 1.4, in deriving the fluctuation-dissipation theorem.) These quantities have a simple interpretation in terms of averages and mean squared fluctuations,

$$\alpha(t) = \langle\mathbf{a};t\rangle, \qquad \sigma(t) = \langle(\mathbf{a}-\langle\mathbf{a};t\rangle)(\mathbf{a}-\langle\mathbf{a};t\rangle);t\rangle. \tag{2.76}$$

The Fourier transform is the exponential of a quadratic function of ξ, and so (see Appendix 2) the inverse transform is a quadratic or Gaussian function of $\mathbf{a}$,

$$G(\mathbf{a},t|\mathbf{a}_0) = (\det 2\pi\sigma(t))^{-1/2}\exp\left(-\frac{1}{2}(\mathbf{a}-e^{t\Theta}\cdot\mathbf{a}_0)\cdot\sigma(t)^{-1}\cdot(\mathbf{a}-e^{t\Theta}\cdot\mathbf{a}_0)\right). \tag{2.77}$$

The coefficient of the exponential is the normalization factor.

Rotational Diffusion

As an illustration of the use of Fokker-Planck equations, we will work out the orientational time correlation function of a planar Brownian rotator. (This was done already, very briefly in section 1.3 using a Langevin equation). The state is specified by an angle θ and by an angular velocity Ω. The rotator has moment of inertia I. Then the equilibrium distribution function is

$$f_{eq}(\Omega, \theta) = \frac{1}{2\pi}\left(\frac{I}{2\pi kT}\right)^{1/2} \exp\left(-\frac{I}{2kT}\Omega^2\right). \tag{2.78}$$

The Fokker-Planck equation corresponding the Langevin equation in section 1.3, with a change from linear velocity to angular velocity, mass to moment of inertia, and position to angle, is

$$\frac{\partial f}{\partial t} = -\Omega \frac{\partial f}{\partial \theta} + \frac{\zeta kT}{I^2} \frac{\partial}{\partial \Omega}\left(\frac{\partial f}{\partial \Omega} + \frac{I}{kT}\Omega f\right) = \mathcal{D}f. \tag{2.79}$$

The orientational time correlation function to be considered here is

$$C_l(t) = \langle e^{-il\theta(0)} e^{il\theta(t)} \rangle_{eq}, \tag{2.80}$$

where l is an integer (periodicity in angle). According to the previous discussion, this may be written

$$C_l(t) = \int d\theta \int d\Omega\, e^{-il\theta} e^{t\mathcal{D}} e^{il\theta} f_{eq}. \tag{2.81}$$

So we want the solution of the Fokker-Planck equation for the special initial condition

$$f(0) = e^{il\theta} f_{eq}. \tag{2.82}$$

The time-dependent solution will have exactly the same angle dependence as the initial distribution,

$$f(\Omega, \theta, t) = e^{il\theta} \frac{1}{2\pi} f_l(\Omega, t). \tag{2.83}$$

Then the time-correlation function is an integral over Ω only,

$$C_l(t) = \int d\Omega f_l(\Omega, t). \tag{2.84}$$

The Fokker-Planck equation can be solved easily, for this initial distribution, by the following trick: make the substitution

$$f_l(\Omega, t) = \exp \Psi(\Omega, t) \tag{2.85}$$

so that the Fokker-Planck equation transforms into

$$\frac{\partial \Psi}{\partial t} = -il\Omega + \frac{\zeta kT}{I^2}\left(\frac{\partial^2 \Psi}{\partial \Omega^2} + \left(\frac{\partial \Psi}{\partial \Omega}\right)^2 + \frac{I}{kT}\Omega\frac{\partial \Psi}{\partial \Omega} + \frac{I}{kT}\right). \tag{2.86}$$

This looks even harder because of the quadratic nonlinearity but is actually quite easy to solve. We look for a solution where the exponent is quadratic,

$$\Psi = a + b\Omega - \frac{I}{2kT}\Omega^2. \tag{2.87}$$

After some cancelations, this leads to the very simple

$$\frac{da}{dt} + \frac{db}{dt}\Omega = -il\Omega + \frac{\zeta kT}{I^2}b^2 - \frac{\zeta}{I}b\Omega. \tag{2.88}$$

Collecting terms, we find ordinary differential equations for a and b,

$$\frac{da}{dt} = \frac{kT}{I\tau}b^2, \quad \frac{db}{dt} = -il - \frac{1}{\tau}b, \tag{2.89}$$

where

$$\tau = \frac{I}{\zeta} \tag{2.90}$$

is the angular velocity relaxation time. Initially $b(0) = 0$, so that $b(t)$ is

$$b(t) = -il\tau(1 - e^{-t/\tau}) \tag{2.91}$$

and then $a(t) - a(0)$ is

$$a(t) - a(0) = -\frac{kT\tau}{I}l^2\left[t - 2\tau(1 - e^{-t/\tau}) + \frac{\tau}{2}(1 - e^{-2t/\tau})\right]. \tag{2.92}$$

When the a and b parts are combined, and the integral over angular velocity is performed,

$$C_l(t) = \int d\Omega \exp \Psi = \exp\left\{a(t) - a(0) + \frac{kT}{2I}b(t)^2\right\}, \tag{2.93}$$

we find

$$C_l(t) = \exp\left\{-\frac{kT}{I}\tau l^2[t - \tau(1 - e^{-t/\tau})]\right\}. \tag{2.94}$$

(This is identical to the expression found using Langevin equations.) At short times and long times this goes as

$$t \ll \tau, \quad C_l(t) \to \exp\left(-\frac{kT}{I}\frac{1}{2}l^2t^2\right)$$

$$t \gg \tau, \quad C_l(t) \to \exp\left(-\frac{kT\tau}{I}l^2t\right). \qquad (2.95)$$

The crossover from one behavior to the other comes at $t = \tau$.

3

Master Equations

3.1 The Golden Rule

The *Golden Rule* is a colloquial name for a standard formula in time-dependent quantum mechanics. It provides a way to calculate the rate of transition between quantum states. It is approximate and has limited applicability; however, it is the starting point for many treatments of rate processes. Later it will be used in the discussion of master equations. This section gives the usual derivation of the Golden Rule. The following section shows how it is used to relate optical absorption spectra to time correlation functions.

We start with an unperturbed Hamiltonian H, its orthonormalized eigenfunctions $|j\rangle$ and its eigenvalues E_j,

$$H|j\rangle = E_j|j\rangle, \quad \langle j|k\rangle = \delta_{jk}. \tag{3.1}$$

Next we introduce a perturbation $V(t)$, which may be time dependent. This perturbation induces transitions between different unperturbed states. Its matrix elements in the unperturbed representation are

$$V_{jk}(t) = \langle j|V(t)|k\rangle, \quad j \neq k, \tag{3.2}$$

and its diagonal elements vanish.

MASTER EQUATIONS

The perturbed quantum state is determined by Schrodinger's equation,

$$\frac{\partial}{\partial t}\Psi(t) = -\frac{i}{\hbar}(H + V(t))\Psi(t). \tag{3.3}$$

We want to solve this with the condition that the initial quantum state is the unperturbed nth state,

$$\Psi(0) = |n\rangle, \tag{3.4}$$

and we want to find the probability that the system is in a different mth unperturbed state at time t. First we expand the actual wave function in the unperturbed states,

$$\Psi(t) = \sum_j a_j(t)|j\rangle. \tag{3.5}$$

We convert Schrodinger's equation to

$$\frac{\partial}{\partial t} a_j(t) = -\frac{i}{\hbar} E_j a_j(t) - \frac{i}{\hbar} \sum_k V_{jk}(t) a_k(t). \tag{3.6}$$

In the initial state, $a_j(0) = \delta_{jn}$. Integration over time leads to the integral equation,

$$a_j(t) = e^{-iE_jt/\hbar}\delta_{jn} - \frac{i}{\hbar}\sum_k \int_0^t ds\, e^{-iE_j(t-s)/\hbar} V_{jk}(s)a_k(s), \tag{3.7}$$

and to find the solution to first order in the perturbation, we substitute the zeroth-order solution in the integral,

$$a_j(t) = e^{-iE_jt/\hbar}\delta_{jn} - \frac{i}{\hbar}\sum_k \int_0^t ds\, e^{-iE_j(t-s)/\hbar} V_{jk}(s)e^{-iE_ks/\hbar}\delta_{kn} + \cdots. \tag{3.8}$$

In particular, the amplitude of the mth state, when we start in the nth state (and $m \neq n$) is

$$a_m(t) = -\frac{i}{\hbar}\int_0^t ds\, e^{-iE_m(t-s)/\hbar} V_{mn}(s)e^{-iE_ns/\hbar} + \cdots. \tag{3.9}$$

In most applications, the perturbation is either independent of time or periodic with frequency ω,

$$V(t) = V \quad \text{or} \quad V(t) = V\cos\omega t. \tag{3.10}$$

For convenience, we abbreviate

50 NONEQUILIBRIUM STATISTICAL MECHANICS

$$\frac{E_m - E_n}{\hbar} = \omega_{mn}. \tag{3.11}$$

When the perturbation is time independent, the amplitude of the mth state is

$$a_m(t) = -\frac{1}{\hbar} e^{-iE_m t/\hbar} V_{mn} \frac{e^{j\omega_{mn}t} - 1}{\omega_{mn}} + \cdots. \tag{3.12}$$

The probability of finding the system in the mth state is

$$P_m(t) = |a_m(t)|^2 + \cdots = \frac{1}{\hbar^2} |V_{mn}|^2 \Delta(t) + \cdots, \tag{3.13}$$

which contains the time-dependent factor $\Delta(t)$,

$$\Delta(t) = \frac{e^{i\omega_{mn}t} - 1}{\omega_{mn}} \frac{e^{-i\omega_{mn}t} - 1}{\omega_{mn}} = \frac{4\sin^2 \frac{\omega_{mn}t}{2}}{\omega_{mn}^2}. \tag{3.14}$$

Regarded as a function of ω_{mn}, this quantity is sharply peaked at $\omega_{mn} = 0$. Figure 3.1.1 shows Δ, at $t = 10$, as a function of ω_{mn}.

The height of the peak is t^2, and the width of the peak is of the order of $2\pi/t$. By using the integral

$$\int_{-\infty}^{\infty} du \frac{\sin^2 u}{u^2} = \pi, \tag{3.15}$$

we find that the area under the curve is $2\pi t$. Then, in the limit of very large t, the peak approaches a delta function, and

$$\Delta(t) \to 2\pi t \delta(\omega_{mn}) = 2\pi t \hbar \delta(E_m - E_n). \tag{3.16}$$

The probability of finding the system in the mth state is

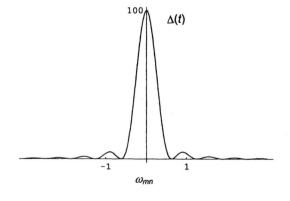

Figure 3.1.1 $\Delta(t)$ as a function of ω_{mn} at time $t = 10$, when the perturbation is time-independent.

$$P_m(t) \to \frac{2\pi}{\hbar} t |V_{mn}|^2 \delta(E_m - E_n) + \cdots . \qquad (3.17)$$

This probability increases linearly with t, and so its time derivative is the rate of transition w_{mn} from state n to state m,

$$w_{mn} = \frac{2\pi}{\hbar} |V_{mn}|^2 \delta(E_m - E_n) + \cdots . \qquad (3.18)$$

This formula is the *Golden Rule*.

When the perturbation is periodic, with nonzero frequency, the amplitude is

$$a_m(t) = -\frac{1}{2\hbar} e^{-iE_m t/\hbar} V_{mn} \left(\frac{e^{i(\omega_{mn}+\omega)t} - 1}{\omega_{mn} + \omega} + \frac{e^{i(\omega_{mn}-\omega)t} - 1}{\omega_{mn} - \omega} \right) + \cdots . \qquad (3.19)$$

In this case, the time-dependent factor $\Delta(t)$ is

$$\Delta(t) = \frac{1}{4} \left| \frac{e^{i(\omega_{mn}+\omega)t} - 1}{\omega_{mn} + \omega} + \frac{e^{i(\omega_{mn}-\omega)t} - 1}{\omega_{mn} - \omega} \right|^2 . \qquad (3.20)$$

For large t, the dependence of $\Delta(t)$ on ω_{mn} is dominated either by ω_{mn} near ω or by ω_{mn} near $-\omega$. This is illustrated in a plot of $\Delta(t)$ as a function of ω_{mn} for the special values $t = 10$ and $\omega = 1$. Figure 3.1.2 clearly shows two peaks at ω_{mn} near 1 and -1.

As in the earlier discussion, the peaks have a height of the order of t^2 and a width of the order of $1/t$ and consequently an area of the order of t. The two peaks can be treated as independent as long as the width of a peak is much smaller than the separation between the two peaks. This requires that $1/t$ should be much smaller than ω, which is true at large enough t as long as ω is not zero. Of

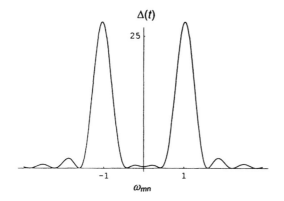

Figure 3.1.2 $\Delta(t)$ as a function of ω_{mn} at time $t = 10$, when the perturbation is periodic with frequency $\omega = 1$.

course, when $\omega = 0$ there can be only one peak, and this condition can never be satisfied. As t increases, the peaks get higher and narrower, and Δ approaches the sum of two delta functions. In the limit of large t,

$$\Delta(t) \to \frac{t\pi}{2}\delta(\omega_{mn} + \omega) + \frac{t\pi}{2}\delta(\omega_{mn} - \omega). \tag{3.21}$$

The rate of transition from state n to state m is given by

$$w_{mn} = \frac{\pi}{2\hbar}|V_{mn}|^2(\delta(E_m - E_n + \hbar\omega) + \delta(E_m - E_n - \hbar\omega)). \tag{3.22}$$

Earlier we saw that if the perturbation is independent of time, the transition rate is

$$w_{mn} = \frac{2\pi}{\hbar}|V_{mn}|^2 \delta(E_m - E_n). \tag{3.23}$$

Note the lack of continuity as ω goes to zero; the two formulas differ by a factor of 2. When the two peaks of Δ are separate, they behave essentially independently at large t. As the two peaks approach each other, there is a constructive interference that doubles the area.

The occurrence of a delta function in the Golden Rule formula always seems strange at first. It suggests that only transitions between states of exactly the same unperturbed energy are allowed and that these transitions have an infinite rate. But we must remember that the delta function appears only as an approximation for the long-time behavior of the function $\Delta(t)$. Further, in normal use it is always associated with sums or integrals over quasi-continuous distributions of states. In standard quantum mechanical use, for example in calculating scattering cross senctions, these sums are taken over final states. However, in statistical mechanical applications, for example in calculating transition rates of a system coupled to a heat bath, the sums are more commonly taken over initial states. This example will be discussed in section 3.3, which deals with master equations.

This standard treatment clearly has flaws. In the first place, it uses first-order perturbation theory. If we replace V by λV, then λ must be small. Second, while the delta functions come from large t, the time cannot really be taken as large as we like. The computed probability of finding the system in the mth state at time t increases linearly with t and will become greater than unity at large enough t, which is not allowed.

So not only must the energy spectrum be quasi-continuous, but there must be a double limitation on the sizes of both λ and t. L. Van

Hove (1955) investigated this question by means of an infinite-order perturbation expansion and resummation. He concluded that in the double limit of small λ and large t, the probability looks roughly like

$$P(t) \approx (\,)\lambda^2 t + (\,)(\lambda^2 t)^2 + (\,)(\lambda^2 t)^3 + \cdots. \tag{3.24}$$

The short time behavior (small $\lambda^2 t$) is given by the Golden Rule. The series in $\lambda^2 t$ sums to a decaying exponential, which is why the probability never exceeds 1. The conclusion is that the Golden Rule formula is applicable as long as λ is small, t is large, and the product $\lambda^2 t$ is of order 1. This is called the "Van Hove limit."

3.2 Optical Absorption Coefficient

Many time correlation functions are determined by spectroscopic measurements. For example, the frequency dependence of the optical absorption coefficient of a substance is determined by the time correlation function of its electric dipole moment. The derivation of this connection is an exercise in applying the quantum mechanical Golden Rule.

Repeating some earlier equations, we start with the Hamiltonian H of the unperturbed system, with quantum states $|a\rangle$ and eigenvalues E_a,

$$H|a\rangle = E_a|a\rangle. \tag{3.25}$$

The time-dependent perturbation is periodic with frequency ω,

$$H' = V\cos\omega t. \tag{3.26}$$

We suppose that the interaction matrix $V_{if} = \langle f|V|i\rangle$ has no diagonal elements. The perturbation causes transitions from an initial state i to a final state f. The transition rate w_{fi} from i to f is given by the Golden Rule formula,

$$w_{fi} = \frac{\pi}{2\hbar}|\langle f|V|i\rangle|^2 \{\delta(E_f - E_i - \hbar\omega) + \delta(E_f - E_i + \hbar\omega)\}. \tag{3.27}$$

We focus on the rate of energy absorption $(dE/dt)_{\text{abs}}$, which is the amount of energy transferred per unit time to the system from the applied perturbation. The rate of energy transfer is the energy difference times the transition rate, $(E_f - E_i)w_{fi}$. Since the system starts out in thermal equilibrium, the initial state is taken from a Boltzmann distribution with probability ρ_i. Further, we are concerned with only the energy absorption at the frequency ω, and we do not care what the final state is. So we sum over all initial i with the Boltzmann weight ρ_i and sum over all final f,

$$(dE/dt)_{\text{abs}} = \sum_f \sum_i \rho_i (E_f - E_i) w_{fi}. \quad (3.28)$$

On using the delta functions in the Golden Rule, this becomes

$$(dE/dt)_{\text{abs}} = \frac{\pi}{2\hbar} \hbar\omega \sum_{i,f} \rho_i |\langle f|V|i\rangle|^2 \{\delta(E_f - E_i - \hbar\omega) - \delta(E_f - E_i + \hbar\omega)\}. \quad (3.29)$$

The matrix elements are symmetric in (i, f), so by switching indices in the second term, we can combine the delta functions,

$$(dE/dt)_{\text{abs}} = \frac{\pi}{2} \omega \sum_{i,f} (\rho_i - \rho_f) |\langle f|V|i\rangle|^2 \, \delta(E_f - E_i - \hbar\omega). \quad (3.30)$$

Because ρ is the Boltzmann distribution, we can relate ρ_f to ρ_i,

$$\rho_f = \rho_i e^{-\beta(E_f - E_i)}, \quad (3.31)$$

and because of the delta function, the exponent in this formula can be changed to $\beta\hbar\omega$, leading to

$$(dE/dt)_{\text{abs}} = \frac{\pi}{2} \omega (1 - e^{-\beta\hbar\omega}) \sum_{i,f} \rho_i |\langle f|V|i\rangle|^2 \, \delta(E_f - E_i - \hbar\omega). \quad (3.32)$$

Now we replace the delta function by its integral representation,

$$\delta(E - \hbar\omega) = \frac{1}{2\pi\hbar} \int_{-\infty}^{\infty} dt \, e^{-i\omega t} e^{itE/\hbar}, \quad (3.33)$$

so that after some rearrangement, the energy absorption is

$$(dE/dt)_{\text{abs}} = \frac{\omega}{4\hbar} (1 - e^{-\beta\hbar\omega}) \int_{-\infty}^{\infty} dt \, e^{-i\omega t} \sum_{i,f} \rho_i \langle i|V|f\rangle e^{itE_f/\hbar} \langle f|V|i\rangle e^{-itE_i/\hbar}. \quad (3.34)$$

Next, we recall that the time dependence of any quantum mechanical operator is given in the Heisenberg representation by

$$V(t) = e^{itH/\hbar} V e^{-itH/\hbar}, \quad (3.35)$$

which has the matrix elements

$$\langle a|V(t)|b\rangle = e^{itE_a/\hbar}\langle a|V|b\rangle e^{-itE_b/\hbar}. \tag{3.36}$$

The energy absorption becomes

$$(dE/dt)_{\text{abs}} = \frac{\omega}{4\hbar}(1 - e^{-\beta\hbar\omega})\int_{-\infty}^{\infty} dt\, e^{-i\omega t} \sum_{i,f} \rho_i \langle i|V|f\rangle\langle f|V(t)|i\rangle. \tag{3.37}$$

The sum over f can be done immediately because the set of all states $\{|j\rangle\}$ is complete,

$$\sum_f |f\rangle\langle f| = 1, \tag{3.38}$$

so that

$$(dE/dt)_{\text{abs}} = \frac{\omega}{4\hbar}(1 - e^{-\beta\hbar\omega})\int_{-\infty}^{\infty} dt\, e^{-i\omega t} \sum_{i} \rho_i \langle i|VV(t)|i\rangle. \tag{3.39}$$

The sum over i produces the thermal equilibrium average,

$$(dE/dt)_{\text{abs}} = \frac{\omega}{4\hbar}(1 - e^{-\beta\hbar\omega})\int_{-\infty}^{\infty} dt\, e^{-i\omega t} \langle VV(t)\rangle_{\text{eq}}. \tag{3.40}$$

The rate of energy absorption is determined by the Fourier transform of the time correlation function of the perturbation V.

Now we can apply this result to the theory of optical absorption. The electric field of an incident light wave has the amplitude $E_0 \cos \omega t$ and is polarized in the direction of the unit vector ε. The electric field interacts with the total electric dipole moment $\mathbf{M}$ of the system; the interaction Hamiltonian is

$$H' = V \cos \omega t, \qquad V = -\mathbf{M} \cdot \varepsilon E_0. \tag{3.41}$$

(Unfortunately, there do not seem to be enough letters in the alphabet; E was used for the energy absorbed, E_0 was used for the electric field, and E_a for the energy of the ath quantum state of the system.) Then the energy absorbed is proportional to the square of the electric field,

$$(dE/dt)_{\text{abs}} = \frac{\omega}{4\hbar}(1 - e^{-\beta\hbar\omega})\int_{-\infty}^{\infty} dt\, e^{-i\omega t} \langle \varepsilon \cdot \mathbf{M}\, \varepsilon \cdot \mathbf{M}(t)\rangle_{\text{eq}} E_0^2. \tag{3.42}$$

If the absorbing system is isotropic, the polarization of the electric field is irrelevant, and the average can be simplified to

$$\langle \varepsilon \cdot \mathbf{M}\, \varepsilon \cdot \mathbf{M}(t)\rangle_{\text{eq}} = \frac{1}{3}\langle \mathbf{M} \cdot \mathbf{M}(t)\rangle_{\text{eq}}. \tag{3.43}$$

To relate this to the experimentally observed absorption coefficient, we divide the energy absorption by the energy flux S of the incoming light wave. This requires an extra calculation in electromagnetic theory, which will not be done here; the energy flux is

$$S = \frac{cn}{8\pi} E_0^2, \qquad (3.44)$$

where c is the velocity of light *in vacuo* and n is the index of refraction. The resulting absorption coefficient $\alpha(\omega) = (dE/dt)_{\text{abs}}/S$ is

$$\alpha(\omega) = \frac{2\pi\omega}{3nc\hbar}(1 - e^{-\beta\hbar\omega}) \int_{-\infty}^{\infty} dt\, e^{-i\omega t} \langle \mathbf{M} \cdot \mathbf{M}(t) \rangle_{\text{eq}}. \qquad (3.45)$$

It is proportional to the Fourier transform of the electric dipole-dipole time correlation function.

While this calculation used quantum mechanics, the classical limit is easy to find. By taking the limit where Planck's constant goes to zero, we get

$$\alpha(\omega) = \frac{2\pi\omega^2\beta}{3nc} \int_{-\infty}^{\infty} dt\, e^{-i\omega t} \langle \mathbf{M} \cdot \mathbf{M}(t) \rangle_{\text{eq}}. \qquad (3.46)$$

This involves the *classical* time correlation function of the total electric dipole moment of the system.

3.3 Quantum Mechanical Master Equations

Master equations describe the dynamics of transitions between states. They look like the equations that describe chemical kinetics; however, concentrations of reactants and products are replaced by probabilities of states, and kinetic rate constants are replaced by transition rates. Further, while chemical kinetic equations can be nonlinear, master equations are inherently linear.

Master equations can be discussed at several levels of abstraction (i.e., distance from reality). The earliest and simplest example is the quantum mechanical Pauli master equation. This will be presented first.

Pauli Master Equation

A quantum mechanical system has the Hamiltonian

$$H = H_0 + \lambda V, \qquad (3.47)$$

where H_0 is an unperturbed Hamiltonian and λV is a perturbation (independent of time). The unperturbed system has eigenstates $|m\rangle$ and eigenvalues E_m,

$$H_0|m\rangle = E_m|m\rangle. \qquad (3.48)$$

MASTER EQUATIONS

The probability that the system is in the mth unperturbed state at time t is $P_m(t)$. (Technically, this is the diagonal element $\rho_{mm}(t)$ of the density matrix in the unperturbed representation—more about this in section 6.1.) The probability of occupation of the mth unperturbed state will change with time because this state is not an eigenstate of the perturbed system. However, if the strength of the perturbation is small, then an unperturbed state can be a good approximation to the corresponding perturbed state. We suppose that all diagonal elements of the perturbation have been included in the unperturbed Hamiltonian, so that the perturbation is strictly off-diagonal, $V_{mm} = 0$.

The Pauli master equation is a gain-loss equation for the probability of occupation of a state,

$$\frac{d}{dt}P_m(t) = \sum_n W_{mn} P_n(t) - \sum_n W_{mn} P_m(t). \tag{3.49}$$

The first term on the right is the rate of gain in state m due to transitions from other states n; the second term is the rate of loss from state m due to transitions to other states. The transition rates W_{mn} are given by the Golden Rule,

$$W_{mn} = \frac{2\pi}{\hbar} \lambda^2 |V_{mn}|^2 \delta(E_n - E_m). \tag{3.50}$$

The Pauli master equation has a microcanonical character. Transitions occur only between states that have almost the same total (unperturbed) energy. The transition rates are symmetric in states,

$$W_{mn} = W_{nm}. \tag{3.51}$$

This is often called "microscopic reversibility." At microcanonical equilibrium, all $P_m(\text{eq})$ have the same value $1/g$, where g is the degeneracy of the unperturbed energy.

Heat Bath Master Equation

But most applications of master equations have a canonical character. For example, the original Hamiltonian may describe the behavior of a system that is weakly coupled to a heat bath, and we may want to know the behavior of the system while the heat bath remains in thermal equilibrium. This leads to a "heat bath" master equation.

The Hamiltonian is now taken to be

$$H = H_s + H_b + \lambda V, \tag{3.52}$$

where H_s operates on unperturbed system states (labeled with Roman letters) and H_b operates on unperturbed heat bath states (labeled with Greek letters),

$$H_s|m\rangle = E_m|m\rangle, \qquad H_b|\alpha\rangle = \varepsilon_\alpha|\alpha\rangle. \tag{3.53}$$

Then the unperturbed product states obey

$$(H_s + H_b)|m\alpha\rangle = (E_m + \varepsilon_\alpha)|m\alpha\rangle. \tag{3.54}$$

The perturbation causes transitions between these product states. The master equation (still "microcanonical") is

$$\frac{d}{dt}P_{m\alpha} = \sum_{n\beta} W_{m\alpha,n\beta} P_{n\beta} - \sum_{n\beta} W_{n\beta,m\alpha} P_{m\alpha}. \tag{3.55}$$

The Golden Rule transition rates are

$$W_{m\alpha,n\beta} = \frac{2\pi}{\hbar}\lambda^2 |\langle m\alpha|V|n\beta\rangle|^2 \delta(E_n + \varepsilon_\beta - E_m - \varepsilon_\alpha). \tag{3.56}$$

Now we assume without proof that the bath remains in thermal equilibrium no matter what state the system is in. (This can be justified by the methods presented in section 6.5.) Then $P_{m\alpha}$ can be factored into a nonequilibrium probability $P_m(t)$ for the system and a thermal equilibrium probability ρ_α for the bath,

$$P_{m\alpha}(t) \cong P_m(t)\rho_\alpha. \tag{3.57}$$

When this is substituted in the master equation, and a sum over α is taken, we get

$$\frac{d}{dt}P_m = \sum_n \sum_\alpha \sum_\beta W_{m\alpha,n\beta}\rho_\beta P_n - \sum_n \sum_\alpha \sum_\beta W_{n\beta,m\alpha}\rho_\alpha P_m, \tag{3.58}$$

which may be rewritten as

$$\frac{d}{dt}P_m = \sum_n w_{mn} P_n - \sum_n w_{nm} P_m. \tag{3.59}$$

The new transition rates (denoted by the lowercase w) are no longer symmetric,

$$w_{mn} = \sum_\alpha \sum_\beta W_{m\alpha,n\beta}\rho_\beta, \qquad w_{nm} = \sum_\alpha \sum_\beta W_{n\beta,m\alpha}\rho_\alpha. \tag{3.60}$$

This master equation describes the relaxation of the system probability distribution $P_m(t)$ to its thermal equilibrium form $P_m(\text{eq})$ at the temperature determined by the heat bath.

MASTER EQUATIONS

Even though the new transition rates are no longer symmetrical, they are still related. The thermal heat bath distribution ρ_α is proportional to $\exp(-\varepsilon_\alpha/kT)$. Then on taking advantage of the constraint on total energy coming from the delta function in the microcanonical W, we find

$$w_{mn}\exp(-E_n/kT) = w_{nm}\exp(-E_m/kT). \tag{3.61}$$

This relation is often called the "principle of detailed balance." It says that at thermal equilibrium, the rates of the forward and backward transitions between any pair of system states, weighted by the probabilities of the initial and final states, are equal to each other.

Illustration

One particular application of the heat bath master equation comes up often. It involves a special perturbation Hamiltonian that is a product of a function of system variables and a function of bath variables,

$$\lambda V = F(\text{system}) \cdot G(\text{bath}) \tag{3.62}$$

(In NMR, for example, F is a nuclear magnetic moment and G is a fluctuating magnetic field; see section 6.4.) Then the transition matrix elements factor,

$$V_{m\alpha,n\beta} = F_{mn} \cdot G_{\alpha\beta}. \tag{3.63}$$

The thermally averaged transition rate becomes

$$w_{mn} = \frac{2\pi}{\hbar}|F_{mn}|^2 \sum_\alpha \sum_\beta \delta(E_m - E_n + \varepsilon_\alpha - \varepsilon_\beta)|G_{\alpha\beta}|^2 \rho_\beta. \tag{3.64}$$

We replace the delta function by its integral representation and rearrange,

$$w_{mn} = \frac{1}{\hbar^2}|F_{mn}|^2 \int_{-\infty}^{\infty} dt\ \exp(i\omega_{mn}t) \cdot$$
$$\sum_\alpha \sum_\beta \exp\left(\frac{it}{\hbar}\varepsilon_\alpha\right) G_{\alpha\beta} \exp\left(-\frac{it}{\hbar}\varepsilon_\beta\right) G_{\beta\alpha}\rho_\beta. \tag{3.65}$$

where $\omega_{mn} = (E_m - E_n)/\hbar$. Now we use the Heisenberg representation of the time-dependent $G(t)$,

$$(G(t))_{\alpha\beta} = \exp\left(\frac{it}{\hbar}\varepsilon_\alpha\right) G_{\alpha\beta} \exp\left(-\frac{it}{\hbar}\varepsilon_\beta\right) \tag{3.66}$$

60 NONEQUILIBRIUM STATISTICAL MECHANICS

The sum over α leads to the matrix $[G(0)G(t)]_{\beta\beta}$, and the sum over β gives the equilibrium average of this quantity,

$$W_{mn} = \frac{1}{\hbar^2}|F_{mn}|^2 \int_{-\infty}^{\infty} dt\ \exp(i\omega_{mn}t)\langle G(0)G(t)\rangle_{eq}. \tag{3.67}$$

The transition rate is proportional to the spectral density of a heat bath time correlation function, $\langle G(0)G(t)\rangle_{eq}$, evaluated at the frequency ω_{mn} of the transition.

This can be applied to the harmonic oscillator heat bath Hamiltonian that was used in section 1.6 to derive a Langevin equation for Brownian motion,

$$H_s = H_s(p,x) \tag{3.68}$$

$$H_B = \sum_j \left(\frac{p_j^2}{2} + \frac{1}{2}\omega_j^2\left(q_j - \frac{\gamma}{\omega_j^2}x\right)^2 \right). \tag{3.69}$$

The perturbation is bilinear in the system coordinate x and the heat bath coordinates q_j. Then the two factors F and G are

$$F = x, \quad G = -\sum \gamma_j q_j. \tag{3.70}$$

Because the heat bath consists of harmonic oscillators, the time correlation function of G is simply

$$\langle G(0)G(t)\rangle_{eq} = \sum_j \gamma_j^2 \cos(\omega_j t)\langle q_j^2\rangle_{eq} + \sum_j \frac{\gamma_j^2}{\omega_j}\sin(\omega_j t)\langle p_j q_j\rangle_{eq}. \tag{3.71}$$

The equilibrium averages can be done quantum mechanically, and there are occasions when this might be necessary, but the results are simpler if we consider only a classical heat bath. Then we find

$$\langle G(0)G(t)\rangle_{eq} = kT\sum_j \frac{\gamma_j^2}{\omega_j^2}\cos(\omega_j t) = kTK(t). \tag{3.72}$$

This contains the same memory function that appeared in the earlier derivation of the Langevin equation. If the heat bath gives rise to Markovian friction, $K(t) = 2\zeta\delta(t)$, then the final result for the transition rate is

$$W_{mn} = \frac{kT\zeta}{\hbar^2}|F_{mn}|^2. \tag{3.73}$$

If the system is also a harmonic oscillator, with mass M and frequency Ω, then matrix elements of F connect adjacent states only. The rate for the downward transition from n to $n-1$ is

$$w_{n-1,n} = \frac{kT\zeta}{2M\Omega} n. \qquad (3.74)$$

The rate for the upward transition is determined by the detailed balance condition,

$$w_{n,n-1} = e^{-\Theta} w_{n-1,n}; \qquad \Theta = \hbar\Omega/kT. \qquad (3.75)$$

While the preceding derivation was based on a classical harmonic oscillator heat bath, considerably more-general situations can be handled in the same way. An old example, (L. Landau and E. Teller, 1936) deals with inelastic energy transfer between a molecular harmonic oscillator and a gas of inert bath molecules. The time correlation function (determined now by the details of molecular collisions) is different, but the matrix elements of F are the same. The only significant change is in the numerical coefficient of n in eq. (3.74).

The resulting heat bath master equation for a harmonic oscillator is

$$\frac{d}{dt} p_n = k(n e^{-\Theta} p_{n-1} + (n+1) p_{n+1} - (n + (n+1)e^{-\Theta}) p_n), \qquad (3.76)$$

where k is a rate constant. A simple exercise is to calculate the relaxation of the average energy of the oscillator. Note that the probabilities are normalized to unity, and that the average energy is

$$\langle E \rangle = \sum_n \hbar\Omega \left(n + \frac{1}{2} \right) p_n. \qquad (3.77)$$

Using the master equation, and rearranging the sums, we find

$$\frac{d}{dt} \langle E \rangle = k \left((1 + e^{-\Theta}) \frac{\hbar\omega}{2} - (1 - e^{-\Theta}) \langle E \rangle \right). \qquad (3.78)$$

For any initial distribution, the average energy decays exponentially to equilibrium (H. Bethe, E. Teller, 1941). Equation (3.76) can actually be solved in full generality for any initial condition (E. Montroll and K. Shuler, 1957).

3.4 Other Kinds of Master Equations

Abstract Master Equations

Master equations are often used without any reference to underlying dynamical models. In general, one must have (1) a set of states, labeled by an index n, (2) probabilities of occupation of

62   NONEQUILIBRIUM STATISTICAL MECHANICS

these states, denoted by $P_n(t)$, and (3) a gain-loss equation with specified transition rates W_{mn} as in the Pauli master equation. The states do not have to be quantum states, and the rates do not have to come from a quantum mechanical calculation. However, certain requirements must be imposed.

The master equation may be written in two equivalent ways, either as a gain-loss equation or as a matrix or operator equation,

$$\frac{d}{dt}P_m(t) = \sum_n W_{mn} P_n(t) - \sum_n W_{nm} P_m$$
$$= \sum_n \mathcal{D}_{mn} P_n(t). \qquad (3.79)$$

The matrix $\mathcal{D}$ is defined by

$$\mathcal{D}_{mn} = (1 - \delta_{mn})W_{mn} - \delta_{mn}\sum_k W_{kn}. \qquad (3.80)$$

In order to conserve probability, this matrix must satisfy the sum rule

$$\sum_m \mathcal{D}_{mn} = 0. \qquad (3.81)$$

Further, its off-diagonal elements must be positive or zero, because transition rates cannot be negative. Then, according to standard matrix theorems, $\mathcal{D}$ has at least one zero eigenvalue, and all its other eigenvalues have negative real parts. These describe an approach to equilibrium. An eigenvector associated with a zero eigenvalue is an equilibrium state. If the matrix is "ergodic," which means that any state can be reached from any other state by a sequence of allowed transitions, then there is only one equilibrium state. If $\mathcal{D}$ has more than one equilibrium state, the process is not ergodic, and different initial states can lead to different stationary states at infinite time.

In a chapter on Fokker-Planck equations, there was a brief discussion of the use of operator methods to deal with averages. These methods can be applied without any essential changes to master equations. In fact, the matrix $\mathcal{D}$ was introduced in order to make the similarities clear. First, the distribution function $f(\mathbf{a}, t)$ is replaced by the probability vector $P_m(t)$. Then integration over $\mathbf{a}$ is replaced by summation over m. The average of any property A_m determined by the state m is

$$\langle A, t \rangle = \sum A_m P_m(t). \qquad (3.82)$$

This average can be obtained either by following the evolution of the probability vector $P_m(t)$,

$$P_m(t) = \sum_n (e^{tD})_{mn} P_n(0) \tag{3.83}$$

or by following the evolution of a *defined* time-dependent property,

$$A_m(t) = \sum_n (e^{tD^\dagger})_{mn} A_n, \tag{3.84}$$

where $D^\dagger$ is the matrix adjoint to the original D. Then the average is

$$\langle A, t \rangle = \sum_m A_m(t) P_m(0). \tag{3.85}$$

Finally, note that abstract master equations may appear as discretizations of Fokker-Planck equations, and that Fokker-Planck equations sometimes appear as continuous approximations to master equations. The main distinction is that Fokker-Planck equations are always parabolic differential equations, having no derivatives higher than the second order, whereas master equations can be much more general.

Random Walks

A common application of master equations is in the treatment of random walks on a lattice. For simplicity, consider an infinite one-dimensional lattice, with sites labeled by j. In this application, by "state" we mean the location of the walker. By "transition," we mean the movement of the walker from j to $j + 1$ or $j - 1$. The probability that the walker is in state j at time t is $P_j(t)$. This satisfies the master equation

$$\frac{d}{dt} P_j = w(P_{j+1} - P_j) + w(P_{j-1} - P_j). \tag{3.86}$$

The rate of change of P_j is the rate w of arrival from $j + 1$ or $j - 1$, less the rate of departure from j.

A typical question is: Given that the walker is at the origin at $t = 0$, what is the probability that he is at site j at time t? To answer this, we first construct the lattice Fourier transform of P_j,

$$g(\theta, t) = \sum_j P_j(t) \exp(i\theta j). \tag{3.87}$$

Then the master equation transforms to

$$\frac{d}{dt} g(\theta, t) = -2w(1 - \cos\theta) g(\theta, t), \tag{3.88}$$

which is easy to solve as an initial value problem,

$$g(\theta, t) = \exp(-2wt(1 - \cos \theta))g(\theta, 0) \tag{3.89}$$

But the initial condition is $P_j(0) = \delta_{j,0}$. Then $g(\theta, 0) = 1$. To find the time-dependent probability, we invert the Fourier transform,

$$P_j(t) = \frac{1}{2\pi} \int_{-\pi}^{\pi} d\theta g(\theta, t) \exp(-i\theta j). \tag{3.90}$$

The resulting integral is a representation of a modified Bessel function $I_j(z)$,

$$I_j(z) = \frac{1}{2\pi} \int_{-\pi}^{\pi} d\theta \exp(z \cos \theta) \exp(-i\theta j), \tag{3.91}$$

so that the solution is

$$P_j(t) = e^{-2wt} I_j(2wt). \tag{3.92}$$

Similar calculations can be done for two- and three-dimensional lattices.

Chemical Kinetics

Master equations are sometimes used to model chemical reaction dynamics. Consider the bimolecular reaction

$$A + B \underset{k_2}{\overset{k_1}{\rightleftarrows}} A + A. \tag{3.93}$$

The number of A molecules is m, and the number of B molecules is n. Because the reaction consists of converting between B and A, the total number $N = m + n$ is conserved.

A state of the system is specified by $[m, n]$. (While the index n is redundant because of conservation, it is helpful to include it explicitly.) Transitions are made only to neighboring states, for example, $[m, n]$ to $[m + 1, n - 1]$, in which a B is converted to an A. The transition rate of this process is expected to be the forward rate constant k_1 times the number of As times the concentration of B,

$$w(m, n \to m+1, n-1) = k_1 m \frac{n}{V}, \tag{3.94}$$

where V is the volume of the system. In the other direction, from $[m, n]$ to $[m - 1, n + 1]$, an A is converted to a B. The transition rate is the backward rate constant k_2 times the number of As times the concentration of A, or

$$w(m, n \to m-1, n+1) = k_2 m \frac{m}{V}. \quad (3.95)$$

This simplistic model of a reaction ignores many important physical questions. However, it is easy to analyze.

The master equation is constructed by accounting for losses from $[m, n]$ to adjoining states and gains to $[m, n]$ from adjoining states. The probability of this state is $P_m(t)$ (recalling that $n = N - m$ is redundant). The resulting equation is

$$\frac{dP_m}{dt} = \frac{k_1}{V}(m-1)(N+1-m)P_{m-1} - \frac{k_1}{V}m(N-m)P_m$$

$$+ \frac{k_2}{V}(m+1)^2 P_{m+1} - \frac{k_2}{V} m^2 P_m. \quad (3.96)$$

Rather than attempting a complete solution of the master equation, we take advantage of the size of the system. New concentration variables are introduced,

$$C = \frac{m}{V}, \quad C_0 = \frac{N}{V}, \quad \rho(C, t) = P_m(t). \quad (3.97)$$

Then in the limit of large V, we can expand in powers of $1/V$, so that, for example,

$$P_{m+1} - P_m \to \frac{1}{V}\frac{\partial \rho}{\partial C}. \quad (3.98)$$

When this expansion is done on all parts of the master equation, and higher orders of $1/V$ are neglected, it leads to a Fokker-Planck equation for ρ,

$$\frac{\partial \rho}{\partial t} = -\frac{\partial}{\partial C}(k_1 C(C_0 - C) - k_2 C^2)\rho$$

$$+ \frac{1}{2V}\frac{\partial^2}{\partial C^2}(k_1 C(C_0 - C) + k_2 C^2)\rho + O\left(\frac{1}{V^2}\right). \quad (3.99)$$

This should be compared with the Fokker-Planck equations discussed earlier; **a** corresponds to C, and the quantities **v** and **B** are

$$v(C) \leftrightarrow k_1 C(C_0 - C) - k_2 C^2,$$
$$B(C) \leftrightarrow k_1 C(C_0 - C) + k_2 C^2. \quad (3.100)$$

In the limit of infinite V, the second derivative term goes away, and the Fokker-Planck equation is like a noise-free Liouville equation. Then any initially sharp distribution $\rho(C, 0) = \delta(C - C(0))$ will remain sharp, and the concentration precisely satisfies the chemical kinetics rate equation,

$$\frac{d}{dt}C(t) = v(C(t)). \tag{3.101}$$

For finite but large V, an initially sharp distribution will broaden in time. Then there are fluctuations about the mean concentration, which are of the order of

$$C(t) - \langle C(t) \rangle = O\left(\frac{1}{\sqrt{V}}\right). \tag{3.102}$$

This calculation shows why deterministic chemical kinetics equations can be used to describe molecular reactions.

4

Reaction Rates

4.1 Transition State Theory

Transition state theory (TST) (E. P. Wigner, 1932), is a way to calculate the rates of chemical reactions, for example, the rearrangement of a molecule A into a different molecule B. It is based on a very simple idea, and sometimes it works. The idea will be presented uncritically in its most elementary form, and some questions about its implementation will be raised.

In this section, the treatment is limited to classical statistical mechanics. The fully quantum mechanical version of TST is complicated by the possibility of tunneling through potential barriers and is still an important research topic.

We consider a system with N degrees of freedom, $j = 1, 2, \ldots, N$, with coordinates x_j and momenta p_j. The first question is, "How do we define a chemical species in this phase space?" For example, we might use a particular "reaction coordinate" x_1 to separate phase space into two regions. The region with negative x_1 is species A, and the region with positive x_1 is species B. All the other coordinates and momenta are collectively denoted by $\mathbf{X}$. The phase space distribution function is $f(p_1, x_1, \mathbf{X}; t)$. The separation is accomplished by using the step function $\Theta(x)$,

$$\Theta(x) = 0 \text{ if } x < 0$$
$$= 1 \text{ if } x > 0. \tag{4.1}$$

We can define the probability P_B of being in region B (or equivalently the concentration of B) as the ensemble average of $\Theta(x_1)$,

$$P_B(t) = \int\int\int dx_1 dp_1 d\mathbf{X} \Theta(x_1) f(p_1, x_1, \mathbf{X}; t) \qquad (4.2)$$

and P_A as the ensemble average of $\Theta(-x_1)$. The time derivative of P_B is determined by operating on f by the Liouville operator and then by taking the adjoint,

$$\frac{d}{dt} P_B(t) = \int\int\int dx_1 dp_1 d\mathbf{X} (L\Theta(x_1)) f(p_1, x_1, \mathbf{X}; t). \qquad (4.3)$$

The effect of L on Θ (the derivative of the step function giving rise to a delta function) is

$$L\Theta(x_1) = \frac{p_1}{m_1} \frac{\partial}{\partial x_1} \Theta(x_1) = \frac{p_1}{m_1} \delta(x_1). \qquad (4.4)$$

This quantity is the product of a velocity p_1/m_1 and a density $\delta(x_1)$ and is therefore a flux density. The integral over p_1 can be split into separate integrals over positive and negative p_1. Negative p_1 corresponds to leaving region B to go to region A, so that the instantaneous rate of *loss* by transitions from B to A is

$$\left(\frac{d}{dt} P_B(t)\right)_{B\to A} = \int_{-\infty}^{0} dp_1 \int d\mathbf{X} \frac{p_1}{m_1} f(p_1, 0, \mathbf{X}; t). \qquad (4.5)$$

The corresponding instantaneous rate of *gain* by transitions from A to B is

$$\left(\frac{d}{dt} P_B(t)\right)_{A\to B} = \int_{0}^{\infty} dp_1 \int d\mathbf{X} \frac{p_1}{m_1} f(p_1, 0, \mathbf{X}; t). \qquad (4.6)$$

So far, this is exact for any x_1 and for any distribution function.

The essential approximation in TST is the assumption that the phase space distribution function in regions A and B maintains a local equilibrium form at all times. This can not really be so; a flow out of a region will surely affect the phase space distribution in that region. For TST to be valid, the contents of a region must relax to equilibrium much faster than the rate of leaving that region, and returns must not be correlated with departures. This separation of time scales is generally hard to justify and is generally contingent on a good choice of phase space regions.

The distribution is determined by the Hamiltonian,

$$H = \sum_{j=1}^{N} \frac{p_j^2}{2m_j} + U(x_1, x_2, \cdots, x_N). \tag{4.7}$$

The full equilibrium distribution is

$$f_{eq} = \frac{1}{Q} e^{-\beta H}, \tag{4.8}$$

where the denominator is the normalization constant,

$$Q = \iiint dx_1 dp_1 d\mathbf{X} e^{-\beta H} = Q_A + Q_B. \tag{4.9}$$

Contributions from the individual regions A and B are

$$Q_A = \iiint_{x_1<0} dp_1 dx_1 d\mathbf{X} e^{-\beta H}, \quad Q_B = \iiint_{x_1>0} dp_1 dx_1 d\mathbf{X} e^{-\beta H}. \tag{4.10}$$

The equilibrium probability of finding the system in region A is $P_A(\text{eq}) = Q_A/Q$, and in region B, $P_B(\text{eq}) = Q_B/Q$. The local equilibrium distribution looks like the full equilibrium distribution, except that it is weighted on either side by the actual amount of A and B that are present at time t, rather than by the equilibrium amount. So, for example, in region B the local equilibrium distribution is

$$f_{B,\text{loc}} = \frac{P_B(t)}{P_B(\text{eq})} f_{eq}. \tag{4.11}$$

When this is inserted in eq. (4.5), one gets the equilibrium flux out of B,

$$\left(\frac{d}{dt} P_B(t)\right)_{B \to A} = \int_{-\infty}^{0} dp_1 \int d\mathbf{X} \frac{p_1}{m_1} f_{eq}(p_1, 0, \mathbf{X}) \frac{P_B(t)}{P_B(\text{eq})}, \tag{4.12}$$

which can be rewritten as a rate equation,

$$\left(\frac{d}{dt} P_B(t)\right)_{B \to A} = -k_{AB} P_B(t). \tag{4.13}$$

The rate constant k_{AB} (after changing signs from $-p_1$ to $+p_1$) is

$$k_{AB} = \int_0^\infty dp_1 \int d\mathbf{X} \frac{p_1}{m_1} f_{eq}(p_1, 0, \mathbf{X}) \frac{1}{P_B(\text{eq})}. \tag{4.14}$$

In the same way, one has

70 NONEQUILIBRIUM STATISTICAL MECHANICS

$$\left(\frac{d}{dt}P_B(t)\right)_{A\to B} = k_{BA}P_A(t), \tag{4.15}$$

where the rate constant k_{BA} is

$$k_{BA} = \int_0^\infty dp_1 \int d\mathbf{X}\, \frac{p_1}{m_1} f_{eq}(p_1, 0, \mathbf{X}) \frac{1}{P_A(\text{eq})}. \tag{4.16}$$

The resulting rate equations are

$$\frac{d}{dt}P_A(t) = -k_{BA}P_A(t) + k_{AB}P_B(t)$$

$$\frac{d}{dt}P_B(t) = -k_{AB}P_B(t) + k_{BA}P_A(t). \tag{4.17}$$

Very often the TST rate constants are written in a form that contains the partition function of the "transition state." This is defined by the Hamiltonian $H^\ddagger$, which is the original H without the momentum p_1 and with x_1 fixed at the dividing boundary $x_1 = 0$. The normalization constant Q is a classical partition function. The corresponding partition function of the transition state is

$$Q^\ddagger = \int d\mathbf{X}\left(e^{-\beta H^\ddagger}\right)_{p_1=0,\,x_1=0} \tag{4.18}$$

Then, after doing the p_1 integration in the numerator of eq. (4.14), the rate constant is

$$k_{AB} = kT\frac{Q^\ddagger}{Q_B}. \tag{4.19}$$

The TST rate constant is the ratio of two partition functions. This formula is often used to argue that there is some kind of thermodynamic equilibrium between the state B and the transition state $\ddagger$; this should not be taken seriously.

An alternative form uses the quantum mechanical partition function rather than the classical one. The only difference here is that the classical limit of the quantum partition function contains an extra factor of Planck's constant h for each degree of freedom, and $Q^\ddagger$ has one fewer degrees of freedom,

$$(Q_B)_{qm} \to \frac{1}{h^N}Q_B, \quad (Q^\ddagger)_{qm} \to \frac{1}{h^{N-1}}Q^\ddagger. \tag{4.20}$$

On changing to the quantum mechanical form, the rate constant becomes

$$k_{AB} = \frac{kT}{h} \frac{(Q^\ddagger)_{qm}}{(Q_B)_{qm}}. \tag{4.21}$$

But this is a cosmetic change only; in the classical limit, Planck's constant cancels out everywhere.

This argument depends crucially on two connected assumptions. One is that the separation between species is well defined by a specific value of x_1. The other is that the contents of regions A and B reach local thermodynamic equilibrium very fast. Generally we don't know a priori whether either assumption is valid. Even if we know experimentally that transitions between A and B follow simple first-order kinetics, this is not ex post facto evidence of rapid equilibration in each region. It is possible that the contents of A come to a nonequilibrium steady state instead of thermal equilibrium, with a different probability of being at the boundary.

In typical applications, the two regions are associated with minima of the potential energy and are separated by a high barrier where the potential energy has a saddle point. Assume that in the neighborhood of the minimum of region B, located at $(b_1, b_2, \ldots, b_N)$, the potential is diagonal,

$$U = U_0 + \frac{1}{2} \sum_i m_i \omega_i^2 (x_i - b_i)^2 + \cdots. \tag{4.22}$$

In the neighborhood of the saddle point, located at $(0, 0, \ldots, 0)$, the potential energy has a maximum in the direction of x_1 and the same minimum location and frequencies in all other directions,

$$U = U_s - \frac{1}{2} a_{11} x_1^2 + \frac{1}{2} \sum_{i=2} m_i \omega_i^2 x_i^2 + \cdots. \tag{4.23}$$

The partition function of region B is

$$Q_B \cong \frac{(2\pi kT)^N}{\prod_{i=1}^{N} \omega_i} \times e^{-\beta U_0}, \tag{4.24}$$

where each degree of freedom gives a factor of kT. The partition function of the transition state (omitting integration over p_1 and x_1) is

$$Q^\ddagger \cong \frac{(2\pi kT)^{N-1}}{\prod_{i=2}^{N} \omega_i} \times e^{-\beta U_s}. \tag{4.25}$$

The resulting rate constant is

$$k_{AB} \cong \frac{\omega_1}{2\pi} e^{-\beta(U_s - U_0)}, \qquad (4.26)$$

which contains the familiar Arrhenius activation energy $U_S - U_0$ and the frequency factor $\omega_1/2\pi$.

The same TST rate constant often appears in a somewhat different context. The time correlation function of the amount of B that is present at time t is defined by

$$C(t) = \langle \Theta(x_1)\Theta(x_1(t)) \rangle_{eq}. \qquad (4.27)$$

Its time derivative is

$$\dot{C}(t) = \langle \Theta(x_1) L \Theta(x_1(t)) \rangle_{eq}$$
$$= -\langle L\Theta(x_1)\Theta(x_1(t)) \rangle_{eq}. \qquad (4.28)$$

Inserting $L\Theta$ leads to

$$\dot{C}(t) = -\left\langle \frac{p_1}{m_1} \delta(x_1)\Theta(x_1(t)) \right\rangle_{eq}. \qquad (4.29)$$

At short times, $x_1(t)$ is $x_1(0) + p_1 t/m_1 + \ldots$, and because of the delta function, the first term can be dropped, so that

$$\dot{C}(t) \to -\left\langle \frac{p_1}{m_1} \delta(x_1)\Theta\left(\frac{p_1}{m_1} t\right) \right\rangle_{eq}. \qquad (4.30)$$

Since t is positive, Θ requires that p_1 is positive. The short time limit is

$$\dot{C}(t \to 0+) = -\int_0^\infty dp_1 \int d\mathbf{X} \frac{p_1}{m_1} \delta(x_1) f_{eq}(p_1, 0, \mathbf{X}). \qquad (4.31)$$

This contains the same integral as in eq. (4.16). The TST rate constant, while based on a questionable assumption about local equilibrium, nevertheless gives the correct initial decay of the time correlation function.

An Example

There is a highly artificial model system for which TST works beautifully. This is the escape of an ideal gas of point particles from a two-dimensional region labeled A, bounded by curved rigid walls, through a very small exit window, labeled w, shown in Fig. 4.1.1.

The particles do not collide with each other, but they do collide elastically with the walls. They move in straight lines between collisions. All particles have the same kinetic energy. A collision cannot change the magnitude $|v|$ of an individual particle's velocity, only its direction.

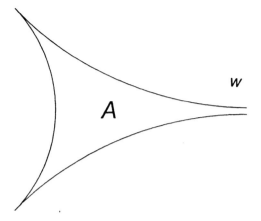

Figure 4.1.1 A model system for which TST works well. An ideal gas of point particles escapes from the region labeled A through the window labeled w.

Because the walls have a negative curvature, initially parallel trajectories will diverge. This means that if all the particles start out with the same vector velocity, but have randomly distributed initial positions, the distribution of velocity directions will be rapidly randomized. If the window is small, the particles will remain in the region for a long time before escaping and consequently they undergo many randomizing collisions. Equilibration within the region is much faster than escape.

It is easy to find the TST escape rate,

$$k_{esc} = \langle \mathbf{v} \cdot \mathbf{n}\, \Theta(\mathbf{v} \cdot \mathbf{n}) \delta(x) \rangle = \frac{2|v|}{2\pi} \frac{w}{A}. \tag{4.32}$$

The exit window is located at $x = 0$, and the unit vector normal to the exit window is $\mathbf{n}$. The first factor is the average velocity of an escaping particle, and the second factor is the average of $\delta(x)$. A is the area of the trapping region, and w is the length of the exit window. Computer simulations of this model (H.-X. Zhou and R. Zwanzig, 1991) confirm that TST works well in the limit of a small window.

4.2 The Kramers Problem and First Passage Times

The Kramers problem (H. A. Kramers, 1940) is to find the rate at which a Brownian particle escapes from a potential well over a potential barrier. One method of attack is based on the theory of first passage times. Since first passage times have other useful applications, they will be discussed first in a general way and then applied to the Kramers problem.

First Passage Times

Suppose that the motion of the set of variables **a** is governed by a Langevin equation. In any single experiment, it follows a specific path **a**(*t*) which wanders through **a**-space. The initial point $\mathbf{a}_0$ starts out somewhere in a "volume" V in this space, bounded by a "surface" ∂V. The first passage time is the first time that the point leaves V. Because of the noise, repeated experiments, even with the same initial position, lead to different paths, and hence different first passage times. The problem here is to find the distribution of first passage times and especially the mean first passage time.

The motion of a cloud of initial points satisfies the Fokker-Planck equation. If we focus on only those points that have not left V by time t, we must remove all paths that have crossed the boundary of V before time t. This can be done by imposing an absorbing boundary condition on ∂V. Then the distribution of points that have not left by time t is $P(\mathbf{a}, t)$, and satisfies

$$\frac{\partial P}{\partial t} = -\nabla_\mathbf{a} \cdot (\mathbf{v}(\mathbf{a})P) + \nabla_\mathbf{a} \cdot \mathbf{B} \cdot \nabla_\mathbf{a} P = \mathcal{D}P,$$

$$P(\mathbf{a}, 0) = \delta(\mathbf{a} - \mathbf{a}_0), \qquad P(\mathbf{a}, t) = 0 \text{ on } \partial V. \qquad (4.33)$$

As before, the Fokker-Planck operator is called $\mathcal{D}$, with the presumption that the boundary condition has been taken into account. The formal operator solution as an initial value problem is

$$P(\mathbf{a}, t) = e^{t\mathcal{D}} \delta(\mathbf{a} - \mathbf{a}_0). \qquad (4.34)$$

Note that P vanishes at long times because of the absorbing boundary condition; eventually all initial points leave. The integral of P over all **a** in the volume V is the number of all starting points that are still in V at time t; it depends on the initial location $\mathbf{a}_0$.

$$S(t, \mathbf{a}_0) = \int_V d\mathbf{a}\, P(\mathbf{a}, t). \qquad (4.35)$$

This also vanishes at long times. The difference $S(t) - S(t + dt)$ is the number of initial points that have not left before time t but have left during the time interval dt following t and therefore determines the distribution of first passage times $\rho(t, \mathbf{a}_0)$,

$$S(t, \mathbf{a}_0) - S(t + dt, \mathbf{a}_0) = \rho(t, \mathbf{a}_0) dt. \qquad (4.36)$$

This provides an explicit way to find ρ,

REACTION RATES 75

$$\rho(t, \mathbf{a}_0) = -\frac{dS(t, \mathbf{a}_0)}{dt}. \tag{4.37}$$

The mean first passage time is the first moment of t,

$$\tau(\mathbf{a}_0) = \int_0^t dt\, t\rho(t, \mathbf{a}_0). \tag{4.38}$$

On using eq. (4.37), integrating by parts, and recalling that S vanishes for large t, one finds

$$\tau(\mathbf{a}_0) = \int_0^\infty dt\, dt\, S(t, \mathbf{a}_0). \tag{4.39}$$

There is a more direct way to calculate ρ, using the operator $\mathcal{D}^\dagger$ that is adjoint to the Fokker-Planck operator $\mathcal{D}$,

$$\tau(\mathbf{a}_0) = \int_0^\infty dt \int d\mathbf{a}\, e^{t\mathcal{D}} \delta(\mathbf{a} - \mathbf{a}_0) = \int_0^\infty dt \int d\mathbf{a}\, \delta(\mathbf{a} - \mathbf{a}_0)\left(e^{t\mathcal{D}^\dagger} 1\right). \tag{4.40}$$

Note that the exponential of the adjoint operates on the number 1. Now the integration over $\mathbf{a}$, with the delta function, replaces $\mathbf{a}$ by $\mathbf{a}_0$ on the right. Then we can drop the subscript "$_0$" and write

$$\tau(\mathbf{a}) = \int_0^\infty dt\, e^{t\mathcal{D}^\dagger} 1; \tag{4.41}$$

the initial location is now $\mathbf{a}$. Next, operate on τ with the adjoint and then do the time integral:

$$\mathcal{D}^\dagger \tau(\mathbf{a}) = \int_0^\infty dt\, \mathcal{D}^\dagger e^{t\mathcal{D}^\dagger} 1 = \int_0^\infty dt\, \frac{d}{dt} e^{t\mathcal{D}^\dagger} 1 = -1. \tag{4.42}$$

Only the lower limit survives; the upper limit vanishes because of the absorbing boundary condition. Then the mean first passage time is determined by solving the inhomogeneous adjoint equation,

$$\mathcal{D}^\dagger \tau(\mathbf{a}) = -1, \qquad \tau(\mathbf{a}) = 0 \text{ on } \partial V. \tag{4.43}$$

The boundary condition in this equation states that any initial point on the boundary will leave immediately; its first passage time is 0.

If $\mathbf{a}$ is the single coordinate x and the Fokker-Planck equation is the one-dimensional Smoluchowski equation,

$$\frac{\partial f}{\partial t} = D \frac{\partial}{\partial x} e^{-U(x)/kT} \frac{\partial}{\partial x} e^{U(x)/kT} f, \tag{4.44}$$

then the calculation of the mean first passage time can be reduced to quadrature. The adjoint equation is

76 NONEQUILIBRIUM STATISTICAL MECHANICS

$$De^{U(x)/kT}\frac{\partial}{\partial x}e^{-U(x)/kT}\frac{\partial}{\partial x}\tau(x)=-1. \quad (4.45)$$

The coordinate x is the starting position of the Brownian particle. The absorbing barrier is located at b, and we assume that there is a reflecting barrier at a, with $a < x < b$. To solve the equation, divide through by $D\exp[U(X)/kT]$, integrate once over x, multiply through by $\exp[U(x)/kT]$, and integrate once more over x, using the boundary conditions at the two limits a and b,

$$e^{-U(x)/kT}\frac{\partial}{\partial x}\tau(x)=-\frac{1}{D}\int_a^x dz\, e^{-U(z)/kT} \quad (4.46)$$

$$\tau(x)=\frac{1}{D}\int_x^b dy\, e^{U(y)/kT}\int_a^y dz\, e^{-U(z)/kT}. \quad (4.47)$$

Only for one-dimensional systems can the solution be found so easily. For higher dimensional systems, it is necessary to solve a partial differential equation. Quite often one has recourse to computer simulations instead.

The Kramers Problem

The Kramers problem is to determine the rate at which a Brownian particle escapes from a potential well. Two typical situations are shown in the following figure, in which a potential is plotted against a coordinate. Figure 4.2.1A might describe a molecular rearrangement, and Fig. 4.2.1B might describe a molecular dissociation.

When the temperature is low (compared with the barrier height), the particle will spend a lot of time near the potential minimum where it started, and only rarely will Brownian motion take it to the top of the barrier. Once there, the particle is equally likely to fall to either

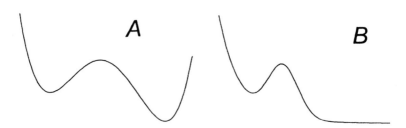

Figure 4.2.1 Potential energy as a function of reaction coordinate in two typical cases, A is a potential with two minima, and B is a potential that allows escape from a single minimum.

side of the barrier. If it goes to the right-hand side, in Fig. 4.2.1A it will fall rapidly to the other minimum, stay there for awhile, and then perhaps cross back to the original minimum. In Fig. 4.2.1B, it will not return.

When the Brownian motion is treated fully, using both position and velocity, the problem is fairly hard and will not be done here. But when the motion is purely diffusive, governed by a Smoluchowski equation, and the barrier is high (or the temperature is low), it is quite easy to find a rate of crossing. The rate of arrival at the barrier is estimated by taking the reciprocal of the first passage time to the barrier. Because a particle at the top of the barrier is equally likely to go either way, the rate of crossing is one half of the rate of arrival. A general formula for mean first passage times was derived in the previous section. The absorbing barrier is placed at the maximum x_{max} of the potential $U(x)$, and $U_{max} = U(x_{max})$. The initial position is x, and the reflecting barrier at $x = a$ is provided by a repelling potential at $x \to -\infty$. Then the mean first passage time from x to x_{max} is

$$\tau(x) = \frac{1}{D}\int_x^{x_{max}} dy\, e^{U(y)/kT}\int_{-\infty}^{y} dz\, e^{-U(z)/kT}. \tag{4.48}$$

When kT is small, the integral over z is dominated by the potential near the minimum,

$$U(z) = U_{min} + \frac{1}{2}\omega_{min}^2 (z - x_{min})^2 + \cdots. \tag{4.49}$$

Then the upper limit of integration can be replaced by infinity, and the integral is

$$\int_{-\infty}^{y} dz\, \exp(-U(z)/kT) \cong \int_{-\infty}^{\infty} dz\, \exp\left(-\frac{U_{min}}{kT}\right)\exp\left(-\frac{\omega_{min}^2}{2kT}(z - x_{min})^2\right)$$

$$= \exp\left(-\frac{U_{min}}{kT}\right)\sqrt{\frac{2\pi kT}{\omega_{min}^2}}. \tag{4.50}$$

The integral over y is dominated by the potential near the barrier and has the quadratic expansion,

$$U(y) = U_{max} - \frac{1}{2}\omega_{max}^2 (y - x_{max})^2 + \cdots. \tag{4.51}$$

The integral over y is practically independent of x as long as x is near the potential minimum, so the lower limit can be replaced by minus infinity,

$$\int_x^{x_{max}} dy \exp\left(\frac{U(y)}{kT}\right) \cong \int_{-\infty}^{x_{max}} dy \exp\left(\frac{U_{max}}{kT}\right) \exp\left(-\frac{\omega_{max}^2}{2kT}(y-x_{max})^2\right)$$

$$= \frac{1}{2} \exp\left(\frac{U_{max}}{kT}\right) \sqrt{\frac{2\pi kT}{\omega_{max}^2}}. \qquad (4.52)$$

The factor 1/2 appears because only half of the Gaussian is included. The mean first passage time (MFPT) (in the high barrier limit) is

$$\tau(x) \cong \frac{1}{2D} \frac{2\pi kT}{\omega_{min}\omega_{max}} \exp\left(\frac{U_{max}-U_{min}}{kT}\right). \qquad (4.53)$$

Recall that D is kT/ζ. The rate of arrival is $1/\tau$, and the rate of crossing, k_K, is half of that, so that

$$k_K \cong \frac{\omega_{min}\omega_{max}}{2\pi\zeta} \exp\left(-\frac{U_{max}-U_{min}}{kT}\right). \qquad (4.54)$$

Referring back to the figure, this is the rate of crossing from left to right in Fig. 4.2.1A; there is a corresponding rate of crossing from right to left, with a different U_{min} and ω_{min}. In Fig. 4.2.1B, this is the rate of escape, since there is no return. The relation of the Kramers rate to the transition state theory rate k_{TST} is simply

$$k_K = \frac{\omega_{max}}{\zeta} k_{TST}. \qquad (4.55)$$

Because this calculation was based on the Smoluchowski equation, the escape rate is correct only in the high friction limit of Brownian motion. The low friction limit is treated in the following section.

4.3 The Kramers Problem and Energy Diffusion

In the preceding section, the Kramers problem was treated by means of a first passage time calculation. Brownian motion over the barrier was treated by a Smoluchowski equation, which means that the results are applicable only in the high friction limit. If the friction is weak, another approach must be taken. This is based on the concept of "energy diffusion."

If there is no friction at all, the particle's energy is conserved, and its motion in the potential well is periodic. Weak noise and friction produce slow random changes of the energy, and this happens in times much longer than a period of oscillation. Eventually, the particle's energy reaches the barrier energy and the particle escapes. In this limit, it is helpful to think in terms of new variables, replacing position and

REACTION RATES

momentum by the action (or energy) and angle of the oscillation. The angular distribution randomizes rapidly, and the energy distribution drifts slowly or "diffuses."

A general method of treating slow dynamics will be presented in section 9.3. At this point, a simple approximate treatment will be given. The starting point is the Fokker-Planck equation for the phase space distribution function $f(x, p; t)$ of a particle in one dimension,

$$\frac{\partial}{\partial t} f = -L_0 f + \zeta k T \frac{\partial}{\partial p} f_{eq} \frac{\partial}{\partial p} \frac{f}{f_{eq}}, \tag{4.56}$$

where L_0 is the Liouville operator,

$$L_0 = \frac{p}{m} \frac{\partial}{\partial x} + F(x) \frac{\partial}{\partial p}, \tag{4.57}$$

f_{eq} is the equilibrium distribution, determined by the Hamiltonian,

$$H = \frac{p^2}{2m} + U(x), \tag{4.58}$$

and the force is $F(x) = -dU/dx$. The distribution of energy can be found from

$$g(E; t) = \iint dx dp \, \delta(H(x, p) - E) f(x, p; t). \tag{4.59}$$

The delta function selects all points on the surface of constant energy E. Because the unperturbed energy is conserved, $L_0 H = 0$, the energy distribution obeys the integrated Fokker-Planck equation,

$$\frac{\partial}{\partial t} g(E; t) = \zeta k T \iint dx dp \, \delta(H - E) \frac{\partial}{\partial p} f_{eq} \frac{\partial}{\partial p} \frac{f}{f_{eq}}. \tag{4.60}$$

The basic approximation is to replace the actual distribution function $f(x, p; t)$ on the right-hand side by a function of the Hamiltonian, so that $f(x, p; t)$ becomes $\phi(H; t)$. This function is determined by requiring that it leads to the correct energy distribution,

$$\iint dx dp \, \delta(H - E) \phi(H; t) = \iint dx dp \, \delta(H - E) f(x, p; t)$$
$$= g(E; t) = \phi(E; t) \iint dx dp \, \delta(H - E). \tag{4.61}$$

The remaining integral is the microcanonical partition function,

$$\Omega(E) = \iint dx dp \, \delta(H - E), \tag{4.62}$$

so that the function ϕ is

$$\phi(E;t) = \frac{g(E;t)}{\Omega(E)}. \tag{4.63}$$

The ratio f/f_{eq} becomes

$$\frac{f(x,p;t)}{f_{eq}(x,p)} \approx \frac{g(H;t)}{g_{eq}(H)}. \tag{4.64}$$

The momentum derivative in eq. (4.60) is

$$\frac{\partial}{\partial p}\frac{f}{f_{eq}} \approx \frac{p}{m}\frac{\partial}{\partial H}\frac{g(H;t)}{g_{eq}(H)} \tag{4.65}$$

leading to

$$\frac{\partial}{\partial t}g(E;t) \approx \zeta kT \iint dxdp\, \delta(H-E)\frac{\partial}{\partial p}f_{eq}\frac{p}{m}\frac{\partial}{\partial H}\frac{g(H;t)}{g_{eq}(H)}. \tag{4.66}$$

Integrating by parts over momentum gives

$$\int dp\, \delta(H-E)\frac{\partial}{\partial p} \to -\int dp\, \frac{\partial}{\partial p}\delta(H-E) \to \frac{\partial}{\partial E}\int dp\, \frac{p}{m}\delta(H-E), \tag{4.67}$$

and eq. (4.60) becomes

$$\frac{\partial}{\partial t}g(E;t) \approx \zeta kT \frac{\partial}{\partial E}\iint dxdp\left(\frac{p}{m}\right)^2 \delta(H-E)f_{eq}(E)\frac{\partial}{\partial E}\frac{g(E;t)}{g_{eq}(E)}. \tag{4.68}$$

This can be rewritten in a "Smoluchowski" form,

$$\frac{\partial}{\partial t}g(E;t) \approx \frac{\partial}{\partial E}D(E)g_{eq}(E)\frac{\partial}{\partial E}\frac{g(E;t)}{g_{eq}(E)}, \tag{4.69}$$

which contains an energy diffusion coefficient,

$$D(E) = \zeta kT\frac{\iint dxdp(p/m)^2 \delta(H-E)}{\iint dxdp\, \delta(H-E)}. \tag{4.70}$$

The momentum integrals can be eliminated by using the identity

$$\int dp\, f(p)\delta(p^2 - a^2) = \frac{f(a)}{2a}; \quad a > 0. \tag{4.71}$$

Then we get

$$D(E) = \frac{2\xi kT}{m} \frac{\int dx \sqrt{E - U(x)}}{\int dx \frac{1}{\sqrt{E - U(x)}}}, \quad (4.72)$$

where the range of integration is over all x such that $E > U(x)$. The integral in the numerator is related to the action defined by an integral around a complete cycle,

$$I(E) = \oint dx\, p(x), \quad p(x) = \sqrt{2m(E - U(x))}. \quad (4.73)$$

The integral in the denominator is related to the derivative dI/dE, which in turn is related to a frequency $\omega(E)$,

$$\left(\frac{\partial I}{\partial E}\right)^{-1} = \frac{\omega(E)}{2\pi}. \quad (4.74)$$

(One must remember that the integrals are over the range of x, where $U(x) < E$ are half of the integrals around a complete cycle.) These definitions provide another way of writing the energy diffusion equation,

$$\frac{\partial}{\partial t} g(E; t) \approx \frac{\partial}{\partial E} \frac{\zeta}{m} I(E) \left[1 + kT \frac{\partial}{\partial E}\right] \frac{\omega(E) g(E; t)}{2\pi}. \quad (4.75)$$

The energy diffusion equation resembles the Smoluchowski equation for spatial diffusion. Position is replaced by energy, and the Boltzmann factor $\exp{-\beta U(x)}$ is replaced by $\Omega(E) \exp{-\beta E}$. The diffusion coefficient now depends on the coordinate E. The essential approximation in deriving this equation is to replace the general x,p dependence of the distribution function by the particular x,p dependence of the Hamiltonian. Another derivation, presented in a later section, shows that this is a good approximation in the weak friction limit.

By exactly the same procedure that was used in the preceding section, we can calculate the rate of escape from a potential well over a barrier. The potential near its minimum is

$$U(x) = \frac{m\omega_0^2}{2} x^2 + \cdots, \quad (4.76)$$

and the barrier energy is E_b. An absorbing boundary condition is used there. The mean first passage time $\tau(E)$ to go from an initial energy E to the barrier is given by the double integral

$$\tau(E) = \int_E^{E_b} dE' \frac{1}{D(E') g_{eq}(E')} \int_0^{E'} dE'' g(E''). \quad (4.77)$$

After inserting expressions for D and g_{eq}, this is

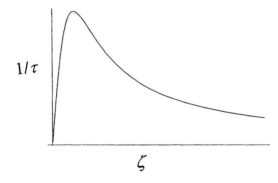

Figure 4.3.1 Escape rate as a function of friction ζ. This is a schematic illustration of the "turnover" from small friction to large friction.

$$\tau(E) = \frac{2m}{\zeta kT} \int_E^{E_b} dE' \frac{e^{\beta E'}}{I(E')} \int_0^{E'} dE'' \Omega(E'') e^{-\beta E''}. \qquad (4.78)$$

Near the potential minimum, I and Ω are given by

$$I(E) = \frac{2m}{\omega_0} E, \qquad \Omega(E) = \frac{\pi}{\omega_0}. \qquad (4.79)$$

At low temperatures, or large β, the second integral is dominated by small E'' and the first integral is dominated by E' near E_b. Then we need Ω at small energy and the action $I(E_b)$ at the barrier energy. Each integration gives a factor $1/\beta$, and the MFPT is

$$\tau \approx \frac{2\pi mkT}{\zeta} \frac{1}{\omega_0 I(E_b)} e^{\beta E_b}. \qquad (4.80)$$

The MFPT in the low friction limit is proportional to $1/\zeta$. The MFPT in the high friction limit, found in the preceding section, is proportional to ζ.

Figure 4.3.1 shows schematically the inverse MFPT $1/\tau$ or escape rate as a function of the friction ζ. There must be a "turnover" connecting the two limits; this was drawn here in a completely arbitrary way. Finding the correct turnover is a difficult problem in mathematical physics, which will not be discussed here.

When there is no possibility of return, the escape rate is $1/\tau$. If, however, the potential has two minima separated by a barrier, particles with energy just over the barrier top will still oscillate back and forth, and only half of the particles are likely to end up on the other side.

5

Kinetic Models

5.1 Kinetic Models

Introduction

Kinetic models originated in attempts to make useful approximations to the Boltzmann equation. However, because of their simplicity and pictorial quality they are often used in entirely different contexts.

The Boltzmann equation was a great achievement of statistical mechanics. It provided a complete and correct treatment of dynamical processes in a gas at low-enough density that only two-body molecular collisions need to be taken into account. It showed how a low-density gas will come to thermal equilibrium at long times (the H-theorem). It is the basis for understanding how the equations of hydrodynamics arise and how the coefficients of viscosity, thermal conductivity, and diffusion are determined by molecular interaction potentials. However, the Boltzmann equation is only applicable to a low-density gas. Further, many of its consequences are quite independent of the details of molecular collisions. For this reason, instead of giving a comprehensive discussion of the Boltzmann equation, a simple "kinetic model" related to the Boltzmann equation will be discussed in a later section.

The state of a system (e.g., one molecule) is given by its position $\mathbf{r}$ and its corresponding conjugate momentum $\mathbf{p}$. In the following treatment, the momentum will be replaced by the velocity $\mathbf{v}$. The number density in the "phase space" specified by position and velocity is $f(\mathbf{r}, \mathbf{v}, t)$. The Boltzmann equation and its kinetic models all have the general form

$$\frac{\partial f}{\partial t} + \mathbf{v}\cdot\nabla_\mathbf{r} f + \frac{1}{m}\mathbf{F}(\mathbf{r})\cdot\nabla_\mathbf{v} f = \left(\frac{\partial f}{\partial t}\right)_{\text{collision}}. \qquad (5.1)$$

The left-hand side of this equation contains the Liouville operator for single particle motion in a potential. $\mathbf{F}(\mathbf{r})$ is an external force acting on the molecule, and m is the molecular mass. For the rest of this section, the external force will be omitted. The right-hand side, usually called the collision integral, accounts for changes in f due to molecular collisions. The gas molecules move freely for awhile; then two molecules collide, and the velocities of both molecules change. The collision integral contains the scattering cross section for a molecular collision, and because collisions require pairs of molecules, it is quadratic in the distribution f. This quadratic nonlinearity is what makes the Boltzmann equation so hard to handle.

In the simplest kinetic model, the collision integral is approximated by

$$\left(\frac{\partial f}{\partial t}\right)_{\text{collision}} = -\lambda(f - f_{\text{eq}}), \qquad (5.2)$$

where λ is a rate constant for the approach of the distribution function to its form at thermal equilibrium. One flaw of eq. (5.2) as a model for gas dynamics is that it leads to incorrect hydrodynamic equations; it violates certain conservation laws. This will be explained more fully later. Better kinetic models for the Boltzmann equation, called BGK models, were developed by P. L. Bhatnagar, E. P. Gross, and M. Krook (1954). These models also ignore details of molecular collisions, replacing the correct collision integral by a simpler approximate form. BGK models lead to hydrodynamic equations that do have the correct general structure, although with incorrect transport coefficients. They will be used in a later chapter to derive the equations of hydrodynamics.

Kinetic Model for Rotational Diffusion

But first, a much simpler kinetic model will be used as an introduction to BGK models. It is intended only as an illustrative example of what can be done, and not as an accurate model of any experimental situation.

This is a kinetic model for the relaxation of angular momentum. It will be used first to derive an equation for rotational diffusion. (The following section contains an application of the same model to the calculation of orientational time correlation functions.) The system is a planar rotator, specified by an angle θ and an angular velocity Ω. Its moment of inertia is I, and the equilibrium distribution of angular velocities is

$$\phi(\Omega) = \left(\frac{I}{2\pi kT}\right)^{1/2} \exp\left(-\frac{I\Omega^2}{2kT}\right). \tag{5.3}$$

There is no external potential, so $U(\theta) = 0$ and $F(\theta) = 0$. The phase space density is $f(\Omega, \theta, t)$. At equilibrium, it has the form

$$f_{eq}(\Omega, \theta) = \frac{1}{2\pi}\phi(\Omega). \tag{5.4}$$

(The 2π comes from the uniform distribution of orientations θ at equilibrium, and the distribution $\varphi(\Omega)$ is normalized to unity.) The kinetic model equation has the same structure as the Boltzmann equation,

$$\frac{\partial f}{\partial t} + \Omega \frac{\partial f}{\partial \theta} = \left(\frac{\partial f}{\partial t}\right)_{collision}. \tag{5.5}$$

The angular velocity of the rotator can change because of random interactions with the environment, which we might call "collisions." The collisions are assumed to be thermalizing. This means that whatever angular velocity the rotator had before a collision, its new angular velocity after a collision is taken randomly from a thermal equilibrium distribution. (In this model, there are no conservation laws, either for total angular momentum or for kinetic energy.) The collisions occur at a rate $1/\tau$. Then we can write a gain-loss kinetic equation (omitting θ and t in f):

$$\frac{\partial}{\partial t}f + \Omega\frac{\partial}{\partial \theta}f = \frac{1}{\tau}\varphi(\Omega)\int d\Omega'\, f(\Omega') - \frac{1}{\tau}f(\Omega). \tag{5.6}$$

In the first term on the right, angular velocities Ω' are removed and thermally distributed angular velocities Ω are gained. The second term on the right accounts for the corresponding losses. The entire right-hand side vanishes at thermal equilibrium.

In the kinetic theory of gases, the Boltzmann equation is used as the starting point for deriving hydrodynamic equations. In this section, the kinetic model equation will be used to derive a rotational diffusion equation. This is an equation for the angle and time dependence of the orientational density ρ that is obtained by integrating out all dependence on angular velocity,

$$\rho(\theta, t) = \int d\Omega\, f(\Omega, \theta, t). \tag{5.7}$$

86   NONEQUILIBRIUM STATISTICAL MECHANICS

One way to get the diffusion equation uses an expansion of the angle dependence of f in Fourier series and the velocity dependence in Hermite polynomials. For convenience, the angular velocity is replaced by

$$v = \left(\frac{I}{2kT}\right)^{1/2} \Omega, \tag{5.8}$$

so that the equilibrium distribution becomes the normalized Gaussian

$$\varphi(v) = \frac{1}{\sqrt{\pi}} \exp - v^2. \tag{5.9}$$

Next, the angle dependence of the distribution is expanded in Fourier components,

$$f(v, \theta, t) = \sum_l f_l(v, t) e^{il\theta}. \tag{5.10}$$

Different Fourier components are uncoupled; the kinetic equation for a single Fourier component is

$$\frac{\partial f_l}{\partial t} + \left(\frac{2kT}{I}\right)^{1/2} vil f_l = \frac{1}{\tau} \varphi(v) \int dv' f_l(v') - \frac{1}{\tau} f_l(v). \tag{5.11}$$

Finally, the velocity dependence of the distribution is expanded in a series of Hermite polynomials,

$$f_l(v, t) = \varphi(v) \sum_m f_{l,m}(t) H_m(v). \tag{5.12}$$

When one uses the recursion formula

$$v H_m(v) = \frac{1}{2} H_{m+1}(v) + m H_{m-1}(v), \tag{5.13}$$

the kinetic equation is transformed to

$$\frac{\partial}{\partial t} f_{l,m} + \left(\frac{2kT}{I}\right)^{1/2} il \left[\frac{1}{2} f_{l,m-1} + (m+1) f_{l,m+1}\right] = -\frac{1}{\tau}(1 - \delta_{m0}) f_{l,m}. \tag{5.14}$$

The equation for $m = 0$ does not involve τ,

$$\frac{\partial}{\partial t} f_{l,0} = -\left(\frac{2kT}{I}\right)^{1/2} il f_{l,1}. \tag{5.15}$$

The time derivatives can be handled by taking Laplace transforms,

$$\hat{f}_l(z) = \int_0^\infty dt\, e^{-zt} f_l(t). \tag{5.16}$$

(Appendix 3 contains a short review of Laplace transforms.) The transform of the time derivative is $z\hat{f}_l(z) - f_l(0)$, which requires that we know the initial values; to make this as simple as possible (at some loss of generality), we assume that initially the velocity dependence has its equilibrium form, but that the angle dependence is arbitrary,

$$f_l(v, 0) = f_{l0}(0)\varphi(v), \qquad f_{lm}(0) = f_{l0}(0)\delta_{m0}. \tag{5.17}$$

Then for $m = 0$, eq. (5.15) becomes

$$z\hat{f}_{l0}(z) = f_{l0}(0) - il\sqrt{\frac{2kT}{I}}\hat{f}_{l1}, \tag{5.18}$$

and for $m = 1, 2, \ldots$, eq. (5.14) gives

$$\left(z + \frac{1}{\tau}\right)\hat{f}_{lm} = -il\sqrt{\frac{2kT}{I}}\left[\frac{1}{2}\hat{f}_{l,m-1} + (m+1)\hat{f}_{l,m+1}\right]. \tag{5.19}$$

This infinite-order three-term recursion problem can be solved in some generality by using continued fractions. (These quantities occur often in nonequilibrium statistical mechanics. Appendix 4 gives a short introduction to continued fractions.) The procedure is to find equations for the ratios of successive terms; in the present case, this is done most easily by introducing the ratios as

$$\hat{K}_{lm} = il(m+1)\sqrt{\frac{2kT}{I}}\frac{\hat{f}_{l,m+1}}{\hat{f}_{lm}}. \tag{5.20}$$

Then eq. (5.18) becomes

$$\hat{f}_{l0} = \frac{f_{l0}(0)}{z + \hat{K}_{l0}(z)}, \tag{5.21}$$

and the recursion formula becomes

$$\hat{K}_{l,m} = (m+1)\frac{kT}{I}l^2\frac{1}{z + \dfrac{1}{\tau} + \hat{K}_{l,m+1}}. \tag{5.22}$$

This procedure generates the continued fraction

$$\hat{f}_{l0} = \frac{f_{l0}(0)}{z+} \frac{l^2 kT/I}{z+\dfrac{1}{\tau}+} \frac{2l^2 kT/I}{z+\dfrac{1}{\tau}+} \cdots. \qquad (5.23)$$

Later a more-compact analytic form will be given.

Evaluating continued fractions can be hard. One strategy is truncation: for example, if we set $\hat{K}_{l1} = 0$, then eq. (5.22) gives an approximation for $\hat{K}_{l0}$, and the continued fraction has two levels,

$$\hat{f}_{l0} \cong \frac{f_{l0}(0)}{z+} \frac{l^2 kT/I}{z+\dfrac{1}{\tau}}. \qquad (5.24)$$

This is a rational function of z, the ratio of a numerator linear in z and a denominator quadratic in z: The inverse Laplace transform can be found from the roots of a quadratic equation and is the sum of two exponentials in t. Or we could set $\hat{K}_{l2} = 0$. Then the continued fraction has one more level, leading to the ratio of a numerator quadratic in z and a denominator cubic in z. The resulting inverse transform is the sum of three exponentials in t. Evidently this is a procedure that one seldom follows for very long.

The Laplace inversion can be easier when one is able to focus only on some limiting case. For example, if τ is very small, then

$$\hat{K}_{l0} = \frac{kTl^2}{I}\tau + O(\tau^2), \qquad (5.25)$$

and the entire continued fraction reduces to

$$\hat{f}_{l0} = \frac{f_{l0}(0)}{z + \dfrac{kTl^2}{I}\tau + O(\tau^2)}. \qquad (5.26)$$

But $f_{l,0}$ is a Fourier component of the angle distribution function $\rho(\theta, t)$, so this result is the solution of the ordinary diffusion equation in angle space,

$$\frac{\partial}{\partial t}\rho(\theta, t) = \frac{kT}{I}\tau \frac{\partial^2}{\partial \theta^2}\rho(\theta, t) + O(\tau^2). \qquad (5.27)$$

This equation is expected to hold only when τ is very small, t is much greater than τ, and deviations from the assumed initial distribution have decayed to zero. These conditions are typical limitations on the validity of diffusion equations.

5.2 Kinetic Models and Rotational Relaxation

In the previous section, a kinetic model was used to derive the rotational diffusion equation in the limit of small τ. One can use the same kinetic model to derive an exact expression for the orientational time correlation function without going through the derivation of a diffusion equation. In fact, the result is precisely the continued fraction that appeared in that derivation. As before, the time correlation function is defined by

$$C_l(t) = \langle e^{-il\theta(0)} e^{il\theta(t)} \rangle_{eq}, \qquad (5.28)$$

where l is an integer. One way to calculate this quantity is to average $\exp(-il\theta)$ with the distribution function that evolves from the initial condition

$$f(\Omega, \theta, 0) = e^{il\theta} f_{eq} = e^{il\theta} \frac{1}{2\pi} \varphi(\Omega). \qquad (5.29)$$

This initial condition, appropriate for the calculation of time correlation functions, is in fact the one that was used as a convenience in the preceding derivation of a diffusion equation. Then the time correlation function is

$$C_l(t) = \int d\theta \int d\Omega\, e^{-il\theta} f(\Omega, \theta, t). \qquad (5.30)$$

The time-dependent solution has exactly the same θ dependence as the initial distribution,

$$f(\Omega, \theta, t) = e^{il\theta} \frac{1}{2\pi} f_l(\Omega, t). \qquad (5.31)$$

Then the time correlation function is an integral over angular velocities only,

$$C_l(t) = \int d\Omega\, f_l(\Omega, t). \qquad (5.32)$$

Because different Fourier components are uncoupled, the kinetic equation for a single component is (as in the earlier derivation of the rotational diffusion equation)

$$\frac{\partial}{\partial t} f_l + il\Omega\, f_l = \frac{1}{\tau} \varphi(\Omega) \int d\Omega'\, f_l(\Omega') - \frac{1}{\tau} f_l(\Omega). \qquad (5.33)$$

The earlier treatment was based on an expansion in Hermite polynomials. A more-compact treatment will be given now.

First we take Laplace transforms in time:

90   NONEQUILIBRIUM STATISTICAL MECHANICS

$$\hat{f}_l(z) = \int_0^\infty dt e^{-zt} f_l(t). \tag{5.34}$$

Then the transformed kinetic equation is

$$z\hat{f}_l - \varphi(\Omega) + il\Omega\,\hat{f}_l = \frac{1}{\tau}\varphi(\Omega)\int d\Omega'\,\hat{f}_l(\Omega') - \frac{1}{\tau}\hat{f}_l. \tag{5.35}$$

By using the abbreviation

$$\hat{g}_l(\Omega) = \frac{1}{z + il\Omega + 1/\tau}, \tag{5.36}$$

we can rearrange the equation to

$$\hat{f}_l(\Omega) = \hat{g}_l(\Omega)\varphi(\Omega) + \hat{g}_l(\Omega)\frac{1}{\tau}\varphi(\Omega)\int d\Omega'\,\hat{f}_l(\Omega'). \tag{5.37}$$

But now we can integrate this over Ω and then solve for the integral,

$$\int d\Omega\,\hat{f}_l(\Omega) = \frac{\int d\Omega\,\hat{g}_l(\Omega)\varphi(\Omega)}{1 - \frac{1}{\tau}\int d\Omega\,\hat{g}_l(\Omega)\varphi(\Omega)}. \tag{5.38}$$

It is convenient to abbreviate again:

$$\hat{G}_l = \int d\Omega\,\hat{g}_l(\Omega)\varphi(\Omega) = \int d\Omega\,\frac{\varphi(\Omega)}{z + il\Omega + 1/\tau}. \tag{5.39}$$

With some effort, the integral can be evaluated analytically; the result is

$$\hat{G}_l(z) = \sqrt{\pi a}\,\exp\!\left(a\!\left(z + \frac{1}{\tau}\right)^2\right)\mathrm{Erfc}\!\left(\sqrt{a}\!\left(z + \frac{1}{\tau}\right)\right), \tag{5.40}$$

where Erfc is the complementary error function and a is

$$a = \frac{I}{2kTl^2}. \tag{5.41}$$

The inverse Laplace transform of this function is easy to find:

$$G_l(t) = \exp\!\left(-\frac{kT}{2I}l^2 t^2 - \frac{t}{\tau}\right). \tag{5.42}$$

Further, the correlation function is the integral of the solution over angular velocity. Then we find for the Laplace transform of the correlation function:

KINETIC MODELS 91

$$\hat{C}_l(t) = \frac{\hat{G}_l}{1 - \hat{G}_l/\tau}. \tag{5.43}$$

The complementary error function Erfc(z) has a continued fraction expansion; when that is used, this Laplace transform is precisely the same continued fraction that was derived earlier. It does not appear that the Laplace transform can be inverted exactly; the results of a numerical calculation will be presented shortly. Also, eq. (5.43) is the solution of a convolution equation in time, which is easy to solve numerically:

$$C_l(t) = G_l(t) + \frac{1}{\tau}\int_0^t ds\, G_l(s) C_l(t-s). \tag{5.44}$$

In the limit of infinite τ, or no interaction with the environment, the time correlation function approaches the ideal rotator limit,

$$C_l(t) \to \exp\left(-\frac{kT}{2I}l^2 t^2\right). \tag{5.45}$$

This behavior was seen earlier.

In the limit of very small τ, it is easiest to work with the Laplace transforms. While this limit was already treated in an earlier section, it may be helpful to look at it again without the complications of a general continued fraction expansion. When τ is very small, the time dependence of $G_l(t)$ is dominated by the factor $\exp(-t/\tau)$; the other Gaussian factor decays much more slowly. Then the Laplace transform is

$$\hat{G}_l(z) = \int_0^\infty dt\, e^{-zt} e^{-t/\tau}\left(1 - \frac{kT}{2I}l^2 t^2 + \cdots\right). \tag{5.46}$$

The integration is done term by term, and then two terms are recombined into a single denominator,

$$\hat{G}_l(z) = \frac{1}{z + 1/\tau} - \frac{kTl^2}{I}\frac{1}{(z + 1/\tau)^3} + \cdots$$

$$\cong \frac{1}{z + 1/\tau + \dfrac{kTl^2}{I}\dfrac{1}{z + 1/\tau} + \cdots}. \tag{5.47}$$

When this is put into eq. (5.43), we immediately get

$$\hat{C}_l(z) \cong \frac{1}{z + \dfrac{kTl^2}{I}\dfrac{1}{z + 1/\tau} + \cdots} \cong \frac{1}{z + \dfrac{kTl^2 \tau}{I}\dfrac{1}{1 + z\tau} + \cdots}. \tag{5.48}$$

This is the same expression that was found by the lowest order truncation of the continued fraction in the treatment of rotational diffusion. For very small τ, the denominator $1 + z\tau$ can be dropped, and

$$\hat{C}_l(z) \cong \frac{1}{z + \dfrac{kTl^2\tau}{I}}. \tag{5.49}$$

The decay of the time correlation function is approximately exponential, with a long lifetime,

$$C_l(z) \cong \exp\left(-\frac{kTl^2\tau}{I}t\right). \tag{5.50}$$

This is exactly the same behavior that was found from the Langevin equation and the Fokker-Planck equation.

The complete time dependence of the time correlation function, for various values of τ, can be obtained by using the Stehfest algorithm (see Appendix 3) to invert the Laplace transform numerically. The time scale was fixed by choosing $I = 2kTl^2$. Figure 5.2.1 shows four curves, using $\tau = 10$ (the lowest curve), 1, 0.5, and 0.1 (the highest curve). The transition from Gaussian to exponential behavior is obvious.

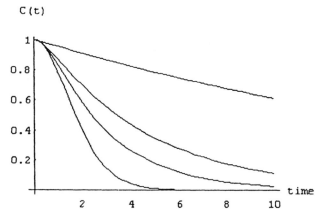

Figure 5.2.1 Results of numerical inversion of the Laplace transform of the orientational time correlation function. The relaxation time τ takes the values 10 (lowest curve), 1, 0.5, and 0.1 (highest curve).

5.3 BGK Equation and the *H*-Theorem

The BGK Equation

Now we turn to the BGK equation as an approximation to the Boltzmann equation, and use it to derive Boltzmann's *H*-theorem. In the next section it is used to derive the equations of hydrodynamics.

First, some definitions, analogous to those used in connection with rotational diffusion, are needed. Earlier, we used a probability distribution function; here, it is more conventional to use a molecular mass density. We define $f(\mathbf{V}, \mathbf{r}, t)$ as the mass per unit volume in the six-dimensional "phase space" determined by the spatial location $\mathbf{r}$ and the molecular velocity $\mathbf{V}$. The distinction between the two forms is one of normalization only—probability distribution functions are necessarily normalized to unity; the density f is normalized to the total mass of the system. The main advantage of this choice, which is conventional in kinetic theory, is that we do not have to keep track of factors of the molecular mass m. The spatial number density is $n(\mathbf{r}, t)$, but it is more helpful in hydrodynamics to use the mass density $\rho(\mathbf{r}, t) = mn(\mathbf{r}, t)$. The fluid velocity is $\mathbf{v}(\mathbf{r}, t)$, and the momentum density is $\rho(\mathbf{r}, t)\mathbf{v}(\mathbf{r}, t)$. The temperature is $T(\mathbf{r}, t)$. We could use, instead, the internal energy density per unit mass, $e(\mathbf{r}, t)$. The mass density and temperature are scalar functions of position and time; the fluid velocity is a vector in three dimensions. These five functions are called the hydrodynamic fields (see Appendix 5 for further discussion). Then the ordinary mass density and the other hydrodynamic fields are related to the phase space mass density by integrals over velocity. The mass density (recall that a factor of m is included in the definition of f) is

$$\rho(\mathbf{r}, t) = \int f(\mathbf{V}, \mathbf{r}, t) d\mathbf{V}. \tag{5.51}$$

The momentum density is

$$\rho(\mathbf{r}, t)\mathbf{v}(\mathbf{r}, t) = \int \mathbf{V} f(\mathbf{V}, \mathbf{r}, t) d\mathbf{V}. \tag{5.52}$$

The molecular velocity $\mathbf{C}$ relative to the bulk motion of the fluid is

$$\mathbf{C}(\mathbf{r}, t) = \mathbf{V} - \mathbf{v}(\mathbf{r}, t), \tag{5.53}$$

and the internal energy density is given by

$$\rho(\mathbf{r}, t)e(\mathbf{r}, t) = \int \frac{1}{2}|\mathbf{C}(\mathbf{r}, t)|^2 f(\mathbf{V}, \mathbf{r}, t) d\mathbf{V}. \tag{5.54}$$

Note that the internal energy is defined by the mean kinetic energy of molecular motion relative to the local fluid velocity. The local temperature T is defined by equating the internal energy per unit volume to $3nkT/2$,

$$\rho(\mathbf{r},t)e(\mathbf{r},t) = \frac{3}{2}n(\mathbf{r},t)kT(\mathbf{r},t). \tag{5.55}$$

We define the *velocity* average of any quantity A by

$$\langle A \rangle = \int d\mathbf{V} A \frac{f(\mathbf{r},\mathbf{V},t)}{\rho(\mathbf{r},t)}, \tag{5.56}$$

which is normalized so that the average of $A = 1$ is 1. Then we can write

$$\mathbf{v}(\mathbf{r},t) = \langle \mathbf{V} \rangle \tag{5.57}$$

$$e(\mathbf{r},t) = \left\langle \frac{1}{2}C^2(\mathbf{r},t) \right\rangle. \tag{5.58}$$

The BGK kinetic model is based on the assumption that the main effect of molecular collisions is to drive the gas to a state of "local equilibrium." This state is defined by a special three-dimensional Gaussian distribution,

$$f_{\text{loc}}(\mathbf{V},\mathbf{r},t) = \rho(\mathbf{r},t)\left(\frac{m}{2\pi kT(\mathbf{r},t)}\right)^{3/2} \exp\left(-\frac{m}{2kT(\mathbf{r},t)}(\mathbf{V}-\mathbf{v}(\mathbf{r},t))^2\right), \tag{5.59}$$

which gives precisely the five hydrodynamic fields.

The local equilibrium distribution is defined so that the local equilibrium hydrodynamic fields are identical to the actual fields,

$$\langle 1 \rangle = \langle 1 \rangle_{\text{loc}} \tag{5.60}$$

$$\langle \mathbf{V} \rangle = \langle \mathbf{V} \rangle_{\text{loc}} \tag{5.61}$$

$$\langle (\mathbf{V}-\mathbf{v})^2 \rangle = \langle (\mathbf{V}-\mathbf{v})^2 \rangle_{\text{loc}}. \tag{5.62}$$

A particular consequence of these identities, used shortly to derive the H-theorem, is

$$\int d\mathbf{V} f \ln f_{\text{loc}} = \int d\mathbf{V} f_{\text{loc}} \ln f_{\text{loc}}. \tag{5.63}$$

KINETIC MODELS 95

The local equilibrium phase space density is a complicated function of the true phase space density; first, f is used to get ρ, $\mathbf{v}$, and T, and then these quantities are used to get the local equilibrium approximation to f. In the BGK kinetic model, the molecular collision integral is replaced by

$$\left(\frac{\partial f}{\partial t}\right)_{\text{collision}} = -\lambda(f - f_{\text{loc}}), \tag{5.64}$$

and the BGK equation is simply

$$\frac{\partial f}{\partial t} + \mathbf{V} \cdot \nabla_r f = -\lambda(f - f_{\text{loc}}). \tag{5.65}$$

The coefficient λ is the rate of approach to local equilibrium; it is a kind of average collision rate, and its inverse $\tau = 1/\lambda$ is an average time between collisions. We expect the approach to local equilibrium to occur on a molecular time scale. In a gas under standard conditions, the duration of a molecular collision is of the order of 10^{-12} s, the time between collisions is of the order of 10^{-9} s, and a macroscopic time might be defined as any time larger than 10^{-4} s. The BGK model does not take molecular collisions into account directly, so the first time is irrelevant. The rate of occurrence of collisions is a good estimate for the rate λ. This is very large compared with the rates of hydrodynamic processes. Thus we will be concerned mainly with the limit of very large λ or very small τ.

Boltzmann's function H is defined by

$$H = \iint d\mathbf{r} d\mathbf{V} f(\mathbf{V}, \mathbf{r}, t) \ln f(\mathbf{V}, \mathbf{r}, t). \tag{5.66}$$

(This differs in an unimportant way from Boltzmann's definition because of the distinction between a mass density and a probability density. If $m = 1$, the distinction goes away.) The rate of change of H is

$$\frac{dH}{dt} = \iint d\mathbf{r} d\mathbf{V} \frac{\partial f}{\partial t}(1 + \ln f). \tag{5.67}$$

The rate of change of f is given by the BGK equation,

$$\frac{dH}{dt} = \iint d\mathbf{r} d\mathbf{V}(-\mathbf{V} \cdot \nabla_r f - \lambda(f - f_{\text{loc}}))(1 + \ln f). \tag{5.68}$$

On partial integration, the streaming term vanishes, and

$$\frac{dH}{dt} = -\lambda \iint d\mathbf{r} d\mathbf{V}(f - f_{\text{loc}}) \ln f. \tag{5.69}$$

NONEQUILIBRIUM STATISTICAL MECHANICS

Equation (5.63) allows us to write this as

$$\frac{dH}{dt} = -\lambda \iint d\mathbf{r}d\mathbf{V}(f - f_{\text{loc}})\ln\frac{f}{f_{\text{loc}}}. \tag{5.70}$$

But the inequality

$$(x-y)\ln\frac{x}{y} \geq 0 \tag{5.71}$$

is generally true for any x and y. Then we have derived the H-theorem,

$$\frac{dH}{dt} \leq 0. \tag{5.72}$$

H can only decrease until it reaches a constant value. When it does, f approaches f_{loc}, and this is a solution of the BGK equation only if all the parameters ρ, $\mathbf{v}$, and T are independent of position. All solutions of the BGK equation approach true thermodynamic equilibrium at long times.

5.4 BGK Equation and Hydrodynamics

Now we use the BGK equation to derive the equations of hydrodynamics. (The treatment of the BGK equation given here is modeled on one used by H. Grad, 1949, to solve the Boltzmann equation.)

The BGK equation is

$$\frac{\partial f}{\partial t} + \mathbf{V} \cdot \nabla_{\mathbf{r}} f = -\lambda(f - f_{\text{loc}}). \tag{5.73}$$

As before, the velocity average of some function A is defined by

$$\langle A \rangle = \int d\mathbf{V} A \frac{f(\mathbf{V},\mathbf{r},t)}{\rho(\mathbf{r},t)}. \tag{5.74}$$

On multiplying the BGK equation by A and integrating over velocity, we obtain an equation of motion for the velocity average of A,

$$\frac{\partial}{\partial t}\rho\langle A\rangle = \int A\frac{\partial f}{\partial t}d\mathbf{V} = \int A\{-\mathbf{V}\cdot\nabla_{\mathbf{r}}f - \lambda(f - f_{\text{loc}})\}d\mathbf{V}, \tag{5.75}$$

which can be rearranged to

$$\frac{\partial}{\partial t}\rho\langle A\rangle = -\nabla_{\mathbf{r}} \cdot \int \mathbf{V}Afd\mathbf{V} + \int (\mathbf{V}\cdot\nabla_{\mathbf{r}}A)fd\mathbf{V} - \lambda\int A(f - f_{\text{loc}})d\mathbf{V}, \tag{5.76}$$

KINETIC MODELS 97

or, using the notation of velocity averages,

$$\frac{\partial}{\partial t}\rho\langle A\rangle = -\nabla_r \cdot \rho\langle \mathbf{V}A\rangle + \rho\langle \mathbf{V}\cdot\nabla_r A\rangle - \lambda\rho(\langle A\rangle - \langle A\rangle_{\text{loc}}). \quad (5.77)$$

The local equilibrium distribution was defined so that the local equilibrium hydrodynamic fields are identical to the actual fields,

$$\langle 1\rangle = \langle 1\rangle_{\text{loc}} \quad (5.78)$$

$$\langle \mathbf{V}\rangle = \langle \mathbf{V}\rangle_{\text{loc}} \quad (5.79)$$

$$\langle (\mathbf{V}-\mathbf{v})^2\rangle = \langle (\mathbf{V}-\mathbf{v})^2\rangle_{\text{loc}}. \quad (5.80)$$

Then if A is chosen to be 1, $\mathbf{V}$, or $(\mathbf{V}-\mathbf{v})^2$, there is no contribution from the collision term to the time derivative of the average.

On using $A = 1$, we obtain the equation of mass conservation,

$$\frac{\partial}{\partial t}\rho = -\nabla_r \cdot \rho\mathbf{v}. \quad (5.81)$$

(For simplicity, the dependence on position and time is left implicit.) The collision term has no effect. On using $A = \mathbf{V}$, we obtain the equation of momentum conservation

$$\frac{\partial}{\partial t}\rho v = -\nabla_r \cdot \int \mathbf{V}\mathbf{V} f d\mathbf{V}. \quad (5.82)$$

Again, the collision term has no effect. The second term on the right will be rewritten (using $\mathbf{V} = \mathbf{v} + \mathbf{C}$):

$$\frac{\partial}{\partial t}\rho v = -\nabla_r \cdot \rho\mathbf{v}\mathbf{v} - \nabla_r \cdot \sigma, \quad (5.83)$$

where the *stress tensor* σ is defined by

$$\sigma_{ij} = \int C_i C_j f d\mathbf{V}. \quad (5.84)$$

On using $A = V^2/2$, we obtain the equation of energy conservation,

$$\frac{\partial}{\partial t}\left(\rho e + \frac{1}{2}\rho v^2\right) = -\nabla_r \cdot \int \mathbf{V}\frac{1}{2}V^2 f d\mathbf{V}. \quad (5.85)$$

This can be further rewritten by introducing the *heat current* **q** defined by

$$\mathbf{q}_i = \int C_i \frac{1}{2}C^2 f d\mathbf{V}. \quad (5.86)$$

Then we have

$$\frac{\partial}{\partial t}\left(\rho e + \frac{1}{2}\rho v^2\right) = -\nabla_r \cdot \left[\mathbf{v}\left(\rho e + \frac{1}{2}\rho v^2\right) + (\boldsymbol{\sigma} \cdot \mathbf{v}) + \mathbf{q}\right]. \tag{5.87}$$

At this point, we have obtained the five hydrodynamic equations, but we have not yet obtained the Navier-Stokes form for the stress tensor or the Fourier heat law. The trace of the stress tensor is

$$\sum_j \sigma_{jj} = \rho\langle|\mathbf{C}|^2\rangle = 3nkT = 3P, \tag{5.88}$$

where $P = nkT$ is the local pressure. The viscous part of the stress tensor (that part connected with shear flows) is defined by

$$(\sigma')_{jk} = \sigma_{jk} - P\delta_{jk} = \rho\left(\langle C_j C_k \rangle - \frac{1}{3}\langle|\mathbf{C}|^2\rangle\delta_{jk}\right). \tag{5.89}$$

Note that the local equilibrium velocity average of σ' is zero. This means that the viscous stress is governed entirely by departures from local equilibrium.

Now we put the tensor $\mathbf{A} = \mathbf{VV}$ into the equation of motion, Eq. (5.77), leading to

$$\frac{\partial}{\partial t}\rho\langle\mathbf{VV}\rangle = -\nabla \cdot \rho\langle\mathbf{VVV}\rangle - \lambda\rho\{\langle\mathbf{VV}\rangle - \langle\mathbf{VV}\rangle_{\text{loc}}\}. \tag{5.90}$$

Or, if we use $\mathbf{A} = \mathbf{VVV}$, we get

$$\frac{\partial}{\partial t}\rho\langle\mathbf{VVV}\rangle = -\nabla_r \cdot \rho\langle\mathbf{VVVV}\rangle - \lambda\rho\{\langle\mathbf{VVV}\rangle - \langle\mathbf{VVV}\rangle_{\text{loc}}\}. \tag{5.91}$$

Note the same recursion problem that occurred in the earlier discussion of rotational diffusion. The rate of change of an nth moment involves an $(n + 1)$th moment. If λ is very large, an nth moment will approach its local equilibrium value rapidly. This is the key to handling the recursion.

The second moment is

$$\rho\langle V_i V_j \rangle = v_i v_j + P\delta_{ij} + \sigma'_{ij}, \tag{5.92}$$

and when its local equilibrium value is subtracted,

$$\rho\langle V_i V_j \rangle - \rho\langle V_i V_j \rangle_{\text{loc}} = \sigma'_{ij}. \tag{5.93}$$

The third moment can be expanded in the same way,

$$\rho\langle V_iV_jV_k\rangle = \rho v_iv_jv_k + \rho v_i\langle C_jC_k\rangle + \rho v_j\langle C_iC_k\rangle + \rho v_k\langle C_iC_j\rangle + \rho\langle C_iC_jC_k\rangle. \tag{5.94}$$

The local equilibrium value of the third moment is

$$\rho\langle V_iV_jV_k\rangle_{\text{loc}} = \rho v_iv_jv_k + \frac{\rho kT}{m}\{v_i\delta_{jk} + v_j\delta_{ik} + v_k\delta_{ij}\}. \tag{5.95}$$

Now these results can be used to derive a formula for the shear viscosity. Rather than attempting this in full generality, we investigate the simplest case: (1) a steady state where the density, temperature, and pressure are spatially uniform, and only the fluid velocity depends on position; (2) all nonlinear terms are neglected; and (3) the third moment can be replaced by its local equilibrium value. Then in the steady state, $\nabla\cdot\mathbf{v} = 0$ because $\partial\rho/\partial t = 0$, and the dynamical equation for the second moment reduces to

$$0 = -\frac{\rho kT}{m}\left[\frac{\partial v_i}{\partial r_j} + \frac{\partial v_j}{\partial r_i}\right] - \lambda\sigma'_{ij} \tag{5.96}$$

or, on inserting an extra $\nabla\cdot\mathbf{v} = 0$,

$$\sigma'_{ij} = -\frac{nkT}{\lambda}\left\{\frac{\partial v_i}{\partial r_j} + \frac{\partial v_j}{\partial r_i} - \frac{2}{3}\delta_{ij}\nabla\cdot\mathbf{v}\right\}. \tag{5.97}$$

This has the standard Navier-Stokes form of the viscous stress tensor, and the coefficient of shear viscosity is simply

$$\eta = \frac{nkT}{\lambda}. \tag{5.98}$$

If the collision rate λ is proportional to the density, as in the kinetic theory of gases, then the viscosity becomes independent of density (J. C. Maxwell, 1860). In this kinetic model, there is no coefficient of volume viscosity.

The same procedure can be used to derive a formula for the thermal conductivity. To make this simple, we consider only the steady-state case where the temperature is a function of position, the pressure is constant, and the fluid velocity vanishes everywhere. Then $\mathbf{V} = \mathbf{C}$, and the heat current is

$$\mathbf{q} = \frac{\rho}{2}\langle\mathbf{V}V^2\rangle. \tag{5.99}$$

The heat current at local equilibrium vanishes. Then eq. (5.91) for the third moment yields

$$\frac{\partial}{\partial t}\mathbf{q} = 0 = -\nabla \cdot \frac{\rho}{2}\langle \mathbf{V}\mathbf{V}V^2\rangle - \lambda \mathbf{q}. \tag{5.100}$$

We replace the fourth moment by its local equilibrium value,

$$\langle V_i V_j V_k V_k\rangle_{\text{loc}} = \langle V_i V_j\rangle_{\text{loc}}\langle V_k V_k\rangle_{\text{loc}} + 2\langle V_i V_k\rangle_{\text{loc}}\langle V_j V_k\rangle_{\text{loc}}. \tag{5.101}$$

When the local equilibrium second moments are inserted, and the sum over k is performed, we find

$$\langle \mathbf{V}\mathbf{V}V^2\rangle_{\text{loc}} = 5\left(\frac{kT}{m}\right)^2 \mathbf{1}. \tag{5.102}$$

Then the heat current is

$$\mathbf{q} = -\nabla \frac{5}{2\lambda}\rho\left(\frac{kT}{m}\right)^2. \tag{5.103}$$

But at constant pressure, $P = nkT$ is independent of location, so that a factor ρkT can be moved through the gradient operator, leading to

$$\mathbf{q} = -\frac{5}{2\lambda}P\frac{k}{m}\nabla T = -\frac{5}{2\lambda}\frac{nk^2 T}{m}\nabla T, \tag{5.104}$$

which has the form of Fourier's heat law. The thermal conductivity is

$$\kappa = \frac{5nk^2 T}{2m\lambda}. \tag{5.105}$$

In the BGK kinetic model, the ratio of thermal conductivity to viscosity has the universal value

$$\left(\frac{\kappa}{\eta}\right)_{\text{BGK}} = \frac{5}{2}\frac{k}{m}. \tag{5.106}$$

When the Boltzmann equation is used to derive expressions for shear viscosity and thermal conductivity, the corresponding ratio differs by a factor of 3/2,

$$\left(\frac{\kappa}{\eta}\right)_{\text{BE}} = \frac{15}{4}\frac{k}{m}. \tag{5.107}$$

Aside from the question of choosing an appropriate value for λ, the BGK transport coefficients cannot quite correspond to the BE coefficients.

6

Quantum Dynamics

6.1 The Quantum Liouville Operator

For present purposes, the first thing to recall about quantum mechanics is that all dynamical quantities, instead of being functions of the location of a system in phase space as in classical mechanics, are now represented by operators in a Hilbert space of quantum states. In particular, the quantum analog of the classical phase space distribution function is the quantum mechanical density matrix. Since this is seldom discussed in introductory texts on either quantum mechanics or statistical mechanics, it may be helpful to summarize some of its basic properties.

In quantum statistical mechanics, we no longer deal with single or "pure" quantum states, but with "mixed" states to which statistical weights are assigned. Recall how one calculates the equilibrium average of any physical quantity A. First we find the energy eigenvalues E_j and eigenfunctions $\varphi_j(\mathbf{q})$ of the system's Hamiltonian operator, H,

$$H\varphi_j(\mathbf{q}) = E_j \varphi_j(\mathbf{q}), \tag{6.1}$$

where $\mathbf{q}$ denotes the system coordinates. The eigenfunctions can always be orthonormalized so that

$$\int d\mathbf{q}\, \varphi_j^*(\mathbf{q})\, \varphi_k(\mathbf{q}) = \delta_{jk}. \tag{6.2}$$

In the jth energy quantum state, the expectation value of the operator representing A is

$$\langle A \rangle_j = \int d\mathbf{q}\, \varphi_j^*(\mathbf{q})\, A\, \varphi_j(\mathbf{q}). \tag{6.3}$$

The probability that a thermal equilibrium system is in the jth quantum state is

$$\rho_j(\text{eq}) = \frac{1}{Q}\exp(-\beta E_j), \quad Q = \sum_j \exp(-\beta E_j), \tag{6.4}$$

where, as usual, $\beta = 1/kT$ is the inverse temperature. The sum in the partition function Q is over all quantum states, including their degeneracies. The thermal equilibrium average of A is its expectation in the jth quantum state, multiplied by the probability of that state, and then summed over all states,

$$\langle A \rangle_{\text{eq}} = \sum \rho_j(\text{eq}) \langle A \rangle_j. \tag{6.5}$$

But it is not necessary or even desirable to use the representation in which the Hamiltonian is diagonal; after all, one can only find these eigenfunctions and eigenvalues in a few simple cases. The energy eigenfunctions, which form a complete orthonormal set, can be expanded in any other complete orthonormal set of functions $\{f_k(\mathbf{q}), k = 1, 2, 3 \ldots\}$,

$$\int d\mathbf{q}\, f_j^*(\mathbf{q}) f_k(\mathbf{q}) = \delta_{jk}, \tag{6.6}$$

so that

$$\varphi_j(\mathbf{q}) = \sum_k S_{jk} f_k(\mathbf{q}). \tag{6.7}$$

The transformation matrix is

$$S_{jk} = \int d\mathbf{q}\, f_k^*(\mathbf{q})\, \varphi_j(\mathbf{q}). \tag{6.8}$$

The inverse transformation is

$$f_j(\mathbf{q}) = \sum_k S_{kj}^* \varphi_k(\mathbf{q}). \tag{6.9}$$

The transformation matrix $\mathbf{S}$ connecting the two sets of states is unitary,

$$(\mathbf{S}^{-1})_{lk} = S_{kl}^*, \quad \sum_l S_{jl} S_{kl}^* = \delta_{jk}. \tag{6.10}$$

In this representation, the Hamiltonian operator is a matrix that is generally not diagonal, but which can still be expressed in terms of the energy eigenvalues,

QUANTUM DYNAMICS   103

$$H \to H_{jk} = \int d\mathbf{q}\, f_j^*(\mathbf{q}) H f_k(\mathbf{q}) = \sum_m S_{mj} E_m S_{mk}^*. \quad (6.11)$$

The matrix representing the operator $\exp(-\beta H)$, which was diagonal in the energy representation, now is generally not diagonal, and the equilibrium probability distribution becomes an operator or a matrix,

$$\rho(\text{eq}) = \frac{1}{Q} e^{-\beta H} \to \rho_{jk}(\text{eq}) = \sum_m S_{mj} \rho_m(\text{eq}) S_{mk}^*. \quad (6.12)$$

This is the equilibrium *density matrix*.

Any dynamical quantity A has a matrix representation,

$$A \to A_{jk} = \int d\mathbf{q}\, f_j^*(\mathbf{q}) A f_k(\mathbf{q}), \quad (6.13)$$

and the average in a single energy quantum state becomes

$$\langle A \rangle_j = \sum_m \sum_m S_{jm}^* A_{mn} S_{jn}. \quad (6.14)$$

Now the thermal equilibrium average of A is

$$\langle A \rangle_{\text{eq}} = \sum_j \rho_j(\text{eq}) \sum_m \sum_n S_{jm}^* A_{mn} S_{jn}$$

$$= \sum_m \sum_n A_{mn} \left(\sum_j \rho_j(\text{eq}) S_{jm}^* S_{jn} \right)$$

$$= \sum_m \sum_n \left(\sum_j \rho_j(\text{eq}) S_{jm} S_{jn}^* \right) A_{nm}. \quad (6.15)$$

(Note that the indices m,n have been switched.) The sum over j is the equilibrium density matrix. The average becomes

$$\langle A \rangle_{\text{eq}} = \sum_m \sum_n \rho_{mn}(\text{eq}) A_{nm} = \text{Trace } \rho(\text{eq}) \cdot \mathbf{A}. \quad (6.16)$$

The average is the trace of the product of the matrix $\mathbf{A}$ and the matrix $\rho(\text{eq})$. This corresponds to the phase space integral in classical mechanics,

$$(\text{classical}) \int d\mathbf{X}\, A(\mathbf{X}) f(\mathbf{X}) \Leftrightarrow (\text{quantum}) \text{ Trace } \rho \cdot \mathbf{A}. \quad (6.17)$$

Remember that the trace of a matrix is invariant to any orthogonal or unitary transformation and that the trace of a product is invariant to a cyclical permutation of its factors; the order of $\mathbf{A}$ and ρ is not important. Note especially that the partition function Q, defined originally in the energy representation, is a trace and is therefore exactly the same in any other representation.

The Quantum Liouville Operator

Now we turn to quantum dynamics. The evolution of any state is given by the Schrodinger equation,

$$\frac{i}{\hbar}\frac{\partial \varphi}{\partial t} = -H\varphi. \tag{6.18}$$

(For now, we stay with time-independent Hamiltonians.) As an initial value problem, this has the operator solution

$$\varphi(\mathbf{q}, t) = e^{-itH/\hbar}\varphi(\mathbf{q}, 0). \tag{6.19}$$

The expectation value of any dynamical variable A at time t is

$$\langle A;t \rangle = \int d\mathbf{q}\, \varphi^*(\mathbf{q}, t) A \varphi(\mathbf{q}, t) \tag{6.20}$$

$$= \int d\mathbf{q}\, (e^{-itH/\hbar}\varphi(\mathbf{q}, 0))^* A(e^{-itH/\hbar}\varphi(\mathbf{q}, 0)), \tag{6.21}$$

and because H is Hermitian, this can be rearranged to

$$\langle A;t \rangle = \int d\mathbf{q}\, \varphi^*(\mathbf{q}, 0)(e^{itH/\hbar} A e^{-itH/\hbar})\varphi(\mathbf{q}, 0). \tag{6.22}$$

This contains the time-dependent Heisenberg operator $A(t)$, defined by

$$A(t) = e^{itH/\hbar} A e^{-itH/\hbar}. \tag{6.23}$$

This is the quantity in quantum mechanics that corresponds to the time-dependent dynamical variable $A(t)$ in classical mechanics.

The initial rate of change of $A(t)$ contains the commutator of A with the Hamiltonian,

$$\left(\frac{\partial}{\partial t} A(t)\right)_{t=0} = \frac{i}{\hbar}[H, A], \tag{6.24}$$

and by analogy with the corresponding discussion in classical mechanics, the right-hand side will be rewritten as

$$\frac{i}{\hbar}[H, A] = LA, \tag{6.25}$$

which defines the *quantum Liouville operator* L. The classical Poisson bracket has been replaced by the quantum commutator. This is an operator (sometimes called a "superoperator") that works on other operators rather than on quantum states. Thus it turns a matrix with two subscripts into a new matrix with two subscripts and is specified by four subscripts (or a "tetradic"),

QUANTUM DYNAMICS 105

$$(LA)_{jk} = \sum_{m,n} L_{jk,mn} A_{mn}. \tag{6.26}$$

The explicit form of L can be found by evaluating the commutator,

$$L_{jk,mn} = \frac{i}{\hbar}(H_{jm}\delta_{nk} - \delta_{jm}H_{nk}). \tag{6.27}$$

The initial rate of change of A is LA; the initial second derivative is LLA, and so on. Then, as in classical mechanics, we can construct the formal Taylor's series in time and sum it to get

$$A(t) = A + tLA + \frac{1}{2}t^2L^2A + \ldots = e^{tL}A. \tag{6.28}$$

This has a tetradic representation,

$$A_{jk}(t) = \sum_{m,n}(e^{tL})_{jk,mn} A_{mn}. \tag{6.29}$$

On using the Heisenberg representation of $A(t)$, it is easy to find the explicit tetradic form of the Liouville propagator,

$$(e^{tL})_{jk,mn} = (e^{itH/\hbar})_{jm}(e^{-itH/\hbar})_{nk}. \tag{6.30}$$

A particularly useful property is

$$\text{Trace } Ae^{tL}B = \text{Trace } Be^{-tL}A, \tag{6.31}$$

which follows from the invariance of the trace to a cyclic permutation. As in classical mechanics, the Liouville operator is anti-self-adjoint.

Normally, quantum mechanics makes use of the two-sided Heisenberg form of time dependence; but in quantum statistical mechanics, there are some advantages in using the one-sided Liouville form. In particular, the dynamical equations $dA(t)/dt = LA(t)$ appear to be formally the same in both classical and quantum mechanics. Only the Liouville operator is different.

As with classical Liouville operators, modifications are required when the Hamiltonian is time dependent. Then L is a function of time, and the exponential operators must be replaced by time-ordered exponentials or by perturbation expansions. The equilibrium density matrix was defined earlier as the way to calculate equilibrium averages of observables. If a system is initially in equilibrium, but then an external time dependent Hamiltonian is switched on, the initial equilibrium density matrix is converted into a new nonequilibrium form; but the density matrix is still used in the same way to calculate averages of observables.

Suppose that a system is represented initially by the density matrix $\rho(0)$. Then, the average of the observable A at time t is

$$\langle A; t \rangle = \text{Trace } A(t) \cdot \rho(0). \tag{6.32}$$

But as noted earlier, the Liouville operator is-anti-self-adjoint, and the average is also given by

$$\langle A; t \rangle = \text{Trace } A e^{-tL} \rho(0) = \text{Trace } A \rho(t), \tag{6.33}$$

which contains the time-dependent density matrix

$$\rho(t) = e^{-tL} \rho(0). \tag{6.34}$$

As in classical mechanics, the average can be found two ways, either by following the evolution of the dynamical variable and averaging over initial conditions or by following the evolution of the initial distribution and averaging at time t. This is the statistical mechanical version of the well-known Heisenberg-Schrodingen duality.

The time-dependent density matrix satisfies the differential equation

$$\frac{\partial}{\partial t} \rho(t) = -L\rho(t) = -\frac{i}{\hbar}[H, \rho(t)]. \tag{6.35}$$

This is the quantum Liouville equation for the density matrix (sometimes called the von Neumann equation). It has the same formal structure as the classical Liouville equation; only the Liouville operator itself is different.

6.2 Electron Transfer Kinetics

This section deals with a theory of electron transfer reactions (R. A. Marcus, 1960). This may be regarded as the quantum analog of the Kramers problem. Standard treatments of the Kramers problem rely on classical mechanics. However, if the "system" is an electron moving in a potential with two minima, quantum mechanics is certainly needed. A simple model of how a classical heat bath affects quantum tunneling is presented here and used to calculate the rate of electron transfer.

The general idea is that a charged molecule, surrounded by a polar medium, interacts with dielectric polarization. For the charge to jump from one site to another, the medium must fluctuate so that energy is conserved. This suggests that we need variables to describe the charge, or "system," and variables to describe the polarization of the environment or "heat bath." We start with the same Hamiltonian that was used earlier in section 1.6. The system variables (representing the electron) are p and x, and the bath variables (representing the polarized environment) are $\{p_j\}$ and $\{q_j\}$. The system Hamiltonian is

$$H_s = \frac{p^2}{2m} + U(x), \tag{6.36}$$

and the heat bath Hamiltonian, as before, is

$$H_B = \sum_j \left(\frac{p_j^2}{2} + \frac{1}{2}\omega_j^2 \left(q_j - \frac{\gamma_j}{\omega_j^2} x \right)^2 \right). \tag{6.37}$$

The Heisenberg equations of motion for all the variables are exactly like the classical equations except that all the variables are time-dependent operators. The same procedure used earlier to derive the Langevin equation can still be used here—it involves the solution of linear equations for the bath operators and is the same for classical and quantum mechanics. The result is a quantum Langevin equation. If the potential function is quadratic in x, this Langevin equation is linear, and one can proceed to a quantum theory of Brownian motion. However, if the potential contains higher powers of the coordinate, as in barrier crossing or tunneling problems, the quantum Langevin equation is extremely hard to handle. Furthermore, there is no analog of the Fokker-Planck equation.

One way to proceed is to approximate the Hamiltonian H_S by a two-state model. Suppose that the potential energy has two minima separated by a barrier. We introduce two quantum states, $|L\rangle$ localized in the left-hand well, with energy E_L, and $|R\rangle$ localized in the right-hand well, with energy E_R. The matrix elements of the system Hamiltonian are $\langle a| H_S |b\rangle$, where a and b are either L or R. Then, the system Hamiltonian is represented by a 2×2 matrix,

$$H_s = \begin{pmatrix} E_L & V \\ V & E_R \end{pmatrix}. \tag{6.38}$$

Tunneling between wells is characterized by an interaction energy V, which is assumed to be very small.

The heat bath Hamiltonian contains the system coordinate x, which is constrained to two values, x_L and x_R. The coordinate operator becomes a 2×2 matrix,

$$x = \begin{pmatrix} x_L & 0 \\ 0 & x_R \end{pmatrix}. \tag{6.39}$$

We assume that off-diagonal elements can be neglected so that the heat bath does not add anything to the tunneling matrix element V. The heat bath Hamiltonian has two forms, H_L or H_R, depending on whether the system coordinate is localized at x_L or x_R,

$$H_{L,R} = \sum \frac{1}{2}p_j^2 + \sum \frac{1}{2}\omega_j^2 \left(q_j - \frac{\gamma_j}{\omega_j^2} x_{L,R} \right)^2. \tag{6.40}$$

Now the total Hamiltonian is

$$H = \begin{pmatrix} H_L + E_L & V \\ V & H_R + E_R \end{pmatrix}. \tag{6.41}$$

Because this Hamiltonian contains both two-state operators and harmonic oscillator operators, it is often referred to as a "spin-boson" Hamiltonian (two states = spin, harmonic oscillator = boson). In general, spin-boson problems are theoretically challenging; here only a highly simplified model is treated approximately.

If the tunneling constant $V = 0$, the Hamiltonian is completely uncoupled, and its states are easily classified. First, there is a two-component spin state, either $(1, 0)$ for $|L\rangle$ or $(0, 1)$ for $|R\rangle$. Then there are states $|\mu L\rangle$ and $|vR\rangle$ of the heat bath Hamiltonian,

$$H_L|\mu L\rangle = \varepsilon_{\mu L}|\mu L\rangle, \qquad H_R|vR\rangle = \varepsilon_{vR}|vR\rangle. \tag{6.42}$$

The unperturbed states of the 2×2 Hamiltonian are $(1, 0)$ $|\mu L\rangle$ and $(0, 1)$ $|vR\rangle$, and the corresponding eigenvalues are $E_{\mu L} = E_L + \varepsilon_{\mu L}$ and $E_{vR} = E_R + \varepsilon_{vR}$.

Transitions between unperturbed states are caused by the perturbation

$$H' = \begin{pmatrix} 0 & V \\ V & 0 \end{pmatrix}; \tag{6.43}$$

its matrix element between $\langle L|$ and $|R\rangle$ is V. The rate of transition between states is given by the Golden Rule formula,

$$w(\mu L \to vR) = \frac{2\pi}{\hbar} |\langle \mu L|V|vR\rangle|^2 \delta(E_{\mu L} - E_{vR})$$

$$= \frac{2\pi}{\hbar} |V|^2 |\langle \mu L|vR\rangle|^2 \delta(\varepsilon_{\mu L} - \varepsilon_{vR} + E_L - E_R). \tag{6.44}$$

Next we sum over all final oscillator states and average over an equilibrium distribution $\rho_{\mu L}$ of initial oscillator states,

$$w(L \to R) = \frac{2\pi}{\hbar} |V|^2 \sum_\mu \rho_{\mu L} \sum_v |\langle \mu L|vR\rangle|^2 \delta(\varepsilon_{\mu L} - \varepsilon_{vR} + E_L - E_R). \tag{6.45}$$

As in earlier uses of the Golden Rule, the sums can be simplified by first writing the delta function as an integral,

$$\delta(E) = \frac{1}{2\pi\hbar}\int_{-\infty}^{\infty} dt\, e^{itE/\hbar}. \tag{6.46}$$

Then the sums can be rearranged,

$$w(L \to R) = \frac{1}{\hbar^2}|V|^2 \int_{-\infty}^{\infty} dt\, e^{it(E_L - E_R)/\hbar} \sum_\mu \rho_{\mu L} e^{itE_{\mu L}/\hbar} \sum_v |\langle \mu L | vR \rangle|^2 e^{-itE_{vR}/\hbar}. \tag{6.47}$$

The sum over v allows us to replace E_{vR} by H_R,

$$\sum_v \langle \mu L | vR \rangle e^{-itE_{vR}/\hbar} \langle vR | \mu L \rangle = \langle \mu L | e^{-itH_R/\hbar} | \mu L \rangle, \tag{6.48}$$

and in the same way, $E_{\mu L}$ can be replaced by H_L,

$$\sum_\mu \rho_{\mu L} e^{itE_{\mu L}/\hbar} \sum_v |\langle \mu L | vR \rangle|^2 e^{-itE_{vR}/\hbar} = \sum_\mu \rho_{\mu L} \langle \mu L | e^{itH_L/\hbar} e^{-itH_R/\hbar} | \mu L \rangle. \tag{6.49}$$

Finally, the sum over μ is the equilibrium average (over the initial state),

$$w(L \to R) = \frac{1}{\hbar^2}|V|^2 \int_{-\infty}^{\infty} dt\, e^{it(E_L - E_R)/\hbar} \langle e^{itH_L/\hbar} e^{-itH_R/\hbar} \rangle_L. \tag{6.50}$$

In general, the two heat bath Hamiltonians in the exponent do not commute. But if we assume that the heat bath oscillators can be treated as classical, then the operators can be replaced by their classical limits, and these do commute. The difference of the two heat bath Hamiltonians is

$$H_L - H_R = -\sum_j \gamma_j q_j (x_L - x_R) + \frac{K}{2}(x_L^2 - x_R^2), \tag{6.51}$$

where, for convenience, we abbreviate,

$$K = \sum_j \frac{\gamma_j^2}{\omega_j^2}. \tag{6.52}$$

(This is the initial value of the memory function for this model of Brownian motion. It is noteworthy that the time-dependent memory function does not occur here.) The difference is linear in the oscillator coordinate,

$$\langle e^{it(H_L - H_R)/\hbar} \rangle_L = e^{itK(x_R^2 - x_L^2)/2\hbar} \langle e^{-it\sum_j \gamma_j q_j (x_L - x_R)/\hbar} \rangle_L, \tag{6.53}$$

and so it is easy to get the equilibrium average of the exponential. In classical statistical mechanics, each q_j has a Gaussian distribution with the mean value and mean squared deviation,

$$\langle q_j \rangle_L = \frac{\gamma_j}{\omega_j^2} x_L, \quad \langle (q_j - \langle q_j \rangle_L)^2 \rangle_L = \frac{kT}{\omega_j^2}. \tag{6.54}$$

Then, the average of the exponential is

$$\langle e^{it(H_L - H_R)/\hbar} \rangle_L = e^{-itK(x_L - x_R)^2/2\hbar} e^{-\frac{t^2}{2\hbar^2}(x_L - x_R)^2 kTK}. \tag{6.55}$$

The quantity $K(x_L - x_R)^2$ has a special significance. It is connected with the change λ in the heat bath energy ongoing from L to R, or Marcus's reorganization energy,

$$\lambda = \langle H_R - H_L \rangle_L = \frac{K}{2}(x_L - x_R)^2. \tag{6.56}$$

Then the transition rate is

$$w(L \to R) = \frac{|V|^2}{\hbar^2} \int_{-\infty}^{\infty} dt \, e^{it(E_L - E_R)/\hbar} e^{-it\lambda/\hbar} e^{-\lambda k T t^2/\hbar^2}. \tag{6.57}$$

On evaluating the integral over t, we get

$$w(L \to R) = \frac{|V|^2}{\hbar} \frac{\sqrt{\pi}}{\sqrt{\lambda kT}} e^{-\frac{1}{kT}\frac{(E_R - E_L + \lambda)^2}{4\lambda}}, \tag{6.58}$$

which is Marcus's result.

6.3 Two-Level System in a Heat Bath: Dephasing

Many problems can be modeled by a two-state quantum system coupled to a classical heat bath. In this section, we discuss a model in which the system is a molecule that has two quantum states, and the heat bath affects only its energy levels. This leads to a theory of the dephasing of spectral lines.

As an elementary model of dephasing, we use the Hamiltonian

$$\mathcal{H} = \begin{pmatrix} H_B & 0 \\ 0 & H_B + \hbar\omega_0 + \hbar V(B) \end{pmatrix}. \tag{6.59}$$

The heat bath Hamiltonian, treated as classical, is H_B. The energy difference of the two levels is $\hbar\omega_0 + \hbar V(B)$. In the following, we set $\hbar = 1$. The first term is constant; the second term leads to fluctuations in the energy difference due to motions of the heat bath. Note that in this example the Hamiltonian has no off-diagonal terms. The density matrix is

QUANTUM DYNAMICS   111

$$\rho = \begin{pmatrix} \rho_{11} & \rho_{12} \\ \rho_{21} & \rho_{22} \end{pmatrix}. \tag{6.60}$$

The diagonal elements of ρ give the probability of finding the system in either state.

The spectral line shape is determined by the dipole-dipole time correlation function. The molecular dipole moment operator is the matrix of the dipole moment μ for a two-state system. We assume that the molecule has no permanent dipole moment, so that

$$\mu = \begin{pmatrix} 0 & \mu_{12} \\ \mu_{21} & 0 \end{pmatrix}. \tag{6.61}$$

The off-diagonal elements of ρ determine the average transition dipole moment,

$$\langle \mu; t \rangle = \int dB (\mu_{12} \rho_{21}(t) + \mu_{21} \rho_{12}(t)), \tag{6.62}$$

and thus they determine the dipole time correlation function and the spectral line shape. In general, each individual ρ_{ij} is still a matrix in the bath coordinates, but when the heat bath is classical, it can be treated as a function of the bath coordinates and momenta.

The equation of motion for the density matrix (remember $\hbar = 1$) is

$$\frac{\partial}{\partial t} \rho = -i[\mathcal{H}, \rho]. \tag{6.63}$$

The individual elements are uncoupled:

$$\frac{\partial}{\partial t} \rho_{11} = \frac{\partial}{\partial t} \rho_{22} = 0 \tag{6.64}$$

$$\frac{\partial}{\partial t} \rho_{12} = -i[H_B, \rho_{12}] + i(\omega_0 + V)\rho_{12} \tag{6.65}$$

$$\frac{\partial}{\partial t} \rho_{21} = -i[H_B, \rho_{21}] - i(\omega_0 + V)\rho_{21}. \tag{6.66}$$

We assume that the heat bath is classical, so that the commutator can be replaced by the corresponding Poisson bracket or the classical Liouville operator L_B; for example,

$$\frac{\partial}{\partial t} \rho_{12} = -L_B \rho_{12} + i(\omega_0 + V)\rho_{12}. \tag{6.67}$$

In the following, we focus on this particular matrix element. The integral over all bath variables is denoted by $c_{12}(t)$,

$$c_{12}(t) = \int dB \, \rho_{12}(t, B). \tag{6.68}$$

The equilibrium bath distribution is $f(B)$, and

$$L_B f(B) = 0, \quad \int dB \, f(B) = 1. \tag{6.69}$$

As an initial condition, we assume that the bath is initially in equilibrium,

$$\rho_{12}(t=0, B) = f(B), \tag{6.70}$$

so that $c_{12}(0) = 1$. The dipole time correlation function is proportional to the real part of $c_{12}(t)$.

There are two ways to proceed, one exact and the other approximate. First we look at the exact procedure. A new variable $g(t, B)$ is defined by

$$\rho_{12}(t, B) = e^{-L_B t} e^{i\omega_0 t} g(t, B). \tag{6.71}$$

On substituting in eq. (6.65), we obtain

$$\frac{\partial}{\partial t} g(t, B) = iV(t, B) g(t, B), \tag{6.72}$$

where the time-dependent V is

$$V(t, B) = e^{L_B t} V(B) = V(B(t)). \tag{6.73}$$

As with all classical dynamical variables, this depends parametrically on the initial values of B. It is easy to integrate eq. (6.72):

$$g(t, B) = \exp\left(i \int_0^t dt' V(t', B)\right) g(0, B). \tag{6.74}$$

The initial value of g is $f(B)$, and the desired result is

$$c_{12}(t) = e^{i\omega_0 t} \int dB \, \exp\left(i \int_0^t dt' V(t', B)\right) f(B). \tag{6.75}$$

The actual value depends on the specific form of heat bath dynamics. Since this is intended to be a calculation that illustrates methods, we are entitled to choose a simple heat bath. Suppose that V is linear in some quantity that is coupled to a heat bath of harmonic oscillators in the way used in section 1.6 as a model for Brownian motion. Then we can construct a heat bath with appropriate frequencies and coupling constants so that $V(t)$ has a Gaussian distribution with zero mean and with the second moment,

$$\langle V(t) V(t') \rangle = K(t - t') = \gamma e^{-|t-t'|/\tau}. \tag{6.76}$$

QUANTUM DYNAMICS 113

The time integral of $V(t)$ also has a Gaussian distribution. Then the heat bath average of the exponential in eq. (6.75) is

$$\int dB \, \exp\left[i\int_0^t dt' V(t')\right] f(B) = \exp\left[-\frac{1}{2}\int_0^t dt' \int_0^t dt'' \langle V(t')V(t'')\rangle\right]. \tag{6.77}$$

The double time integral can be worked out just as in elementary Brownian motion theory; the result is

$$c_{12}(t) = \exp[i\omega_0 t - \gamma\tau(t - \tau + e^{-t/\tau})]. \tag{6.78}$$

If there are no fluctuations, c_{12} oscillates periodically in t, and the spectral line is sharp. The effect of the fluctuations is to modify the phase of the oscillation, called "dephasing." At long times, $c_{12}(t)$ is exponentially damped, with a rate $\gamma\tau$.

In this example, the heat bath was chosen deliberately to allow an exact solution. Generally we can't do that, and approximate methods are needed. A useful one is as follows. Start with eq. (6.65) and integrate over the bath. The integral of L_B vanishes, and

$$\frac{\partial}{\partial t} c_{12}(t) = i\omega_0 c_{12}(t) + \int dB \, iV(B)\rho_{12}(t, B). \tag{6.79}$$

Next, integrate eq. (6.65) once, as an initial value problem, using the assumed initial value of ρ_{12},

$$\rho_{12}(t) = e^{i\omega_0 t} f(B) + \int_0^t dt' \, e^{(i\omega_0 - L_B)(t-t')} iV(B)\rho_{12}(t', B). \tag{6.80}$$

When this is inserted in the preceding equation, the first term drops out because the first moment of V vanishes,

$$\frac{\partial}{\partial t} c_{12}(t) = i\omega_0 c_{12}(t) + \int dB \, iV(B) \int_0^t dt' \, e^{(i\omega_0 - L_B)(t-t')} iV(B)\rho_{12}(t', B) \tag{6.81}$$

The integral is formally second order in V. Now a crucial assumption, to be verified shortly, is that as long as we want only results to second order, $\rho_{12}(t, B)$ may be replaced by the local equilibrium form,

$$\rho_{12}(t, B) \approx c_{12}(t) f(B) \tag{6.82}$$

in the right-hand side. With this assumption, eq. (6.81) becomes

$$\frac{\partial}{\partial t} c_{12}(t) = i\omega_0 c_{12}(t) - \int_0^t dt' \, e^{i\omega_0(t-t')} K(t-t') c_{12}(t'), \tag{6.83}$$

which contains the same memory function as in eq. (6.76),

$$\int dB V(B) e^{L_B(t-t')} V(B) f(B) = \langle V V(t-t') \rangle = K(t-t'). \tag{6.84}$$

Equation (6.83) can be solved by Laplace transforms,

$$\hat{K}(z) = \int_0^\infty dt\, e^{-zt} K(t) = \frac{\gamma}{z + 1/\tau} \tag{6.85}$$

$$\hat{c}_{12}(z) = \int_0^\infty dt\, e^{-zt} c(t) = \frac{1}{z - i\omega_0 + \hat{K}(z - i\omega_0)}. \tag{6.86}$$

Because the memory function has such a simple form, it is easy to invert the Laplace transform, leading to a sum of two exponentials,

$$c_{12}(t) = e^{i\omega_0 t}\left(a_1 e^{-r_1 t} + a_2 e^{-r_2 t}\right) \tag{6.87}$$

$$r_1 = \frac{1}{2\tau}(1 - \sqrt{1 - 4\gamma\tau^2}), \quad r_2 = \frac{1}{2\tau}(1 + \sqrt{1 - 4\gamma\tau^2}) \tag{6.88}$$

$$a_1 = \frac{1/\tau - r_1}{r_2 - r_1}, \quad a_2 = \frac{1/\tau - r_2}{r_1 - r_2}. \tag{6.89}$$

The approximate solution agrees with the exact solution at short times:

$$c_{12} \to e^{i\omega_0 t}(1 - \gamma t^2/2 + \cdots). \tag{6.90}$$

At long times, and for small γ, they also agree:

$$c_{12} \to (1 + \gamma\tau^2 + \cdots)e^{i\omega_0 t - \gamma\tau t}. \tag{6.91}$$

The approximate solution is based on the local equilibrium assumption in eq. (6.82). This can be justified by the following argument. Define the difference between the correct ρ_{12} and its assumed form $c_{12}f$,

$$\delta\rho = \rho_{12}(t, B) - c_{12}(t)f(B). \tag{6.92}$$

From eq. (6.65), one finds

$$\frac{\partial}{\partial t}\delta\rho = (i\omega_0 - L_B)\delta\rho + iV\rho_{12} - f(B)\int dB iV\rho_{12}. \tag{6.93}$$

This has the operator solution

$$\delta\rho(t) = e^{(i\omega_0 - L_B)t}\delta\rho(0) + \int_0^t dt'\, e^{(i\omega_0 - L_B)(t-t')}\{iV\rho_{12}(t') - f(B)\int dB iV\rho_{12}(t')\}. \tag{6.94}$$

QUANTUM DYNAMICS   115

The first term vanishes; with the initial condition in eq. (6.70), $\delta\rho(0) = 0$. Then $\delta\rho(t)$ is of first order in V and contributes to dc_{12}/dt only to third order in V. The second-order result in eq. (6.83) is unaffected.

Finally, one must be cautious in making Markovian approximations. In eq. (6.83), replacing t' by $t - t'$ leads to

$$\frac{\partial}{\partial t} c_{12}(t) = i\omega_0 c_{12}(t) - \int_0^t dt' e^{i\omega_0 t'} K(t') c_{12}(t - t'). \tag{6.95}$$

The usual Markovian approximation involves dropping the t' in c_{12} and then extending the time integral to infinity,

$$\frac{\partial}{\partial t} c_{12}(t) \cong i\omega_0 c_{12}(t) - \int_0^\infty dt' e^{i\omega_0 t'} K(t') c_{12}(t). \tag{6.96}$$

This is wrong. The reason is that if ω_0 is large, c_{12} is no longer a slow variable. Only the product $\exp(i\omega_0 t) c_{12}(t)$ is slow. The correct Markovian approximation is

$$\frac{\partial}{\partial t} c_{12}(t) \cong i\omega_0 c_{12}(t) - \int_0^\infty dt' K(t') c_{12}(t). \tag{6.97}$$

6.4 Two-Level System in a Heat Bath: Bloch Equations

Another example of a two-level system in a heat bath is a spin-1/2 nucleus in a magnetic field that fluctuates due to motions of the heat bath. Here we derive the Bloch equations for the time dependence of the average magnetization and expressions for the relaxation times T_1 and T_2. (The same derivation can be applied to a two-level molecular system with a more-general Hamiltonian than the one used in the last chapter.)

The nucleus has a magnetic moment that is proportional to its spin angular momentum, $\sigma = (\sigma_x, \sigma_y, \sigma_z)$, which is given explicitly by the Pauli spin matrices,

$$\mathbf{1} = \begin{pmatrix} 1 & 0 \\ 0 & 1 \end{pmatrix}, \quad \sigma_x = \begin{pmatrix} 0 & 1 \\ 1 & 0 \end{pmatrix}, \quad \sigma_y = \begin{pmatrix} 0 & -i \\ i & 0 \end{pmatrix}, \quad \sigma_z = \begin{pmatrix} 1 & 0 \\ 0 & -1 \end{pmatrix}. \tag{6.98}$$

The coefficient of proportionality will be included in the definition of the field strength. The nucleus interacts with a magnetic field having two different sources. One is an externally imposed constant field in the z direction, with strength $\hbar\omega_0/2$, and the other is an environmental fluctuating field $\hbar V/2$. (The extra factor $\hbar/2$ is added to cancel a later factor $2/\hbar$.) Now the two-level Hamiltonian $\mathcal{H}$ is

$$\mathcal{H} = \mathcal{H}_0 + \mathcal{H}_1 \tag{6.99}$$

$$\mathcal{H}_0 = H_B \mathbf{1} - \frac{\hbar \omega_0}{2} \sigma_z \tag{6.100}$$

$$\mathcal{H}_1 = -\frac{\hbar}{2}(V_x \sigma_x + V_y \sigma_y + V_z \sigma_z) = -\frac{\hbar}{2} V \cdot \sigma. \tag{6.101}$$

(A center dot indicates the dot product of two vectors.) The total Hamiltonian is

$$\mathcal{H} = \begin{pmatrix} H_B - \frac{\hbar}{2}(\omega_0 + V_z) & -\frac{\hbar}{2}(V_x - iV_y) \\ -\frac{\hbar}{2}(V_x + iV_y) & H_B + \frac{\hbar}{2}(\omega_0 + V_z) \end{pmatrix}. \tag{6.102}$$

The heat bath Hamiltonian H_B is treated for simplicity as classical, and the bath variables are denoted by B. The fluctuating magnetic field V is a function of B. The equilibrium density matrix is

$$\rho(\text{eq}, B) = \frac{1}{Q} e^{-\beta \mathcal{H}}, \qquad Q = \int dB \text{ trace } e^{-\beta \mathcal{H}}. \tag{6.103}$$

The time-dependent density matrix is $\rho(t, B)$, and throughout we use deviations from equilibrium, for example, $\delta \rho = \rho - \rho(\text{eq})$.

Manipulations of spin matrices are simplified by

$$\sigma_i \sigma_j = \delta_{ij} \mathbf{1} + i \sum_k \varepsilon_{ijk} \sigma_k, \tag{6.104}$$

where $\varepsilon_{ijk} = 1$ if ijk is a cyclic permutation of 123, $\varepsilon_{ijk} = -1$ if ijk is a cyclic permutation of 321, and $\varepsilon_{ijk} = 0$ otherwise. The same quantity appears in the cross product of two vectors,

$$(A \times B)_i = \sum_{jk} \varepsilon_{ijk} A_j B_k. \tag{6.105}$$

Any 2×2 matrix can be expressed as a sum of the four spin matrices. In particular, the deviation of the density matrix from equilibrium is

$$\delta \rho(t, B) = \frac{1}{2}(m_0(t, B)\mathbf{1} + m_x(t, B)\sigma_x + m_y(t, B)\sigma_y + m_z(t, B)\sigma_z). \tag{6.106}$$

QUANTUM DYNAMICS 117

The coefficients m_j are found by taking the trace,

$$m_j(t, B) = \text{trace}\,(\sigma_j \delta\rho(t, B)). \tag{6.107}$$

The average magnetization is obtained by integrating over the bath variables, and its deviation from equilibrium is

$$\langle \delta\sigma_j; t\rangle = \langle \sigma_j; t\rangle - \langle \sigma_j; \text{eq}\rangle = \int dB\, m_j(t, B). \tag{6.108}$$

Because the equilibrium density matrix is independent of time, the deviation from equilibrium $\delta\rho$ satisfies the same Liouville equation as ρ,

$$\frac{\partial}{\partial t}\delta\rho = -\frac{i}{\hbar}[\mathcal{H}, \delta\rho] = -L_B \delta\rho + \frac{i}{2}[\omega_0 \sigma_z, \delta\rho] + \frac{i}{2}[V \cdot \sigma, \delta\rho], \tag{6.109}$$

where L_B is the classical Liouville operator for the bath. When the expansion in eq. (6.106) is put into this, and some spin algebra is worked out, we get four equations for the coefficients m_j:

$$\frac{\partial}{\partial t} m_0 = -L_B m_0 \tag{6.110}$$

$$\frac{\partial}{\partial t} m_x = -L_B m_x + \omega_0 m_y - V_y m_z + V_z m_y \tag{6.111}$$

$$\frac{\partial}{\partial t} m_y = -L_B m_y - \omega_0 m_x - V_z m_x + V_x m_z \tag{6.112}$$

$$\frac{\partial}{\partial t} m_z = -L_B m_z - V_x m_y + V_y m_x. \tag{6.113}$$

It is convenient to use a 3×3 matrix representation,

$$R = \begin{pmatrix} 0 & \omega_0 & 0 \\ -\omega_0 & 0 & 0 \\ 0 & 0 & 0 \end{pmatrix}, \quad W = \begin{pmatrix} 0 & V_z & -V_y \\ -V_z & 0 & V_x \\ V_y & -V_x & 0 \end{pmatrix}, \tag{6.114}$$

so the three-component vector $m = (m_x, m_y, m_z)$ satisfies

$$\frac{\partial}{\partial t} m = -L_B m + R \cdot m + W \cdot m. \tag{6.115}$$

Now we can follow essentially the same procedure as in the preceding chapter. One time integration gives

118   NONEQUILIBRIUM STATISTICAL MECHANICS

$$m(t) = e^{-L_B t}e^{Rt} \cdot m(0) + \int_0^t dt' e^{-L_B(t-t')}e^{R(t-t')} \cdot W \cdot m(t'). \tag{6.116}$$

We need an initial condition. As before, the equilibrium bath distribution is $f(B)$. We restrict the treatment to initial conditions where the density matrix is determined by specified initial deviations of the averages,

$$\delta\rho(0, B) = f(B)\frac{1}{2}\sum_{x,y,z} \langle\delta\sigma_j; 0\rangle\sigma_j. \tag{6.117}$$

The component $m_0(0, B)$ starts at 0 and remains there. The other components are $m_j(0, B) = f(B)\langle\delta\sigma_j; 0\rangle$. Return to eq. (6.109) and integrate over the bath variables. This gives

$$\frac{\partial}{\partial t}\langle\delta\sigma; t\rangle = R \cdot \langle\delta\sigma; t\rangle + \int dB W(B) \cdot m(t, B). \tag{6.118}$$

Now insert eq. (6.116). The initial value term drops out; we require that the bath average of the fluctuating field vanishes:

$$\int dB V_j(B)f(B) = 0, \qquad j = x, y, z. \tag{6.119}$$

The remaining part is

$$\frac{\partial}{\partial t}\langle\delta\sigma; t\rangle = R \cdot \langle\delta\sigma; t\rangle + \int dB W \cdot \int_0^t dt' e^{-L_B(t-t')}e^{R(t-t')} \cdot W \cdot m(t'). \tag{6.120}$$

This equation is still exact. Now we make a local equilibrium approximation in the last term, replacing $m(t)$ by

$$m_j(t, B) \to \langle\delta\sigma_j; t\rangle f(B). \tag{6.121}$$

This leads to an approximate equation for the average spin matrices,

$$\frac{\partial}{\partial t}\langle\delta\sigma; t\rangle \cong R \cdot \langle\delta\sigma; t\rangle + \int_0^t dt' K(t-t') \cdot \langle\delta\sigma; t'\rangle, \tag{6.122}$$

where the memory kernel is

$$K(t) = \int dB W \cdot e^{-L_B t}e^{Rt} \cdot W f(B). \tag{6.123}$$

The classical propagator can be moved to the left, producing $W(t)$, and using $\langle\ \rangle$ to denote a bath average, K becomes a time correlation function,

$$K(t) = \langle W(t) \cdot e^{Rt} \cdot W \rangle. \tag{6.124}$$

In working out the matrix products, we need

$$e^{Rt} = \begin{pmatrix} \cos\omega_0 t & \sin\omega_0 t & 0 \\ -\sin\omega_0 t & \cos\omega_0 t & 0 \\ 0 & 0 & 1 \end{pmatrix}. \tag{6.125}$$

The resulting matrix $K(t)$ contains many terms involving the bath average of a product of Vs. We assume that only the diagonal elements survive the average,

$$\langle V_i(t) V_j \rangle = k_i(t) \delta_{ij}, \tag{6.126}$$

so that

$$K(t) = -\begin{pmatrix} k_y(t) + k_z(t)\cos\omega_0 t & k_z(t)\sin\omega_0 t & 0 \\ -k_z(t)\sin\omega_0 t & k_x(t) + k_z(t)\cos\omega_0 t & 0 \\ 0 & 0 & (k_x(t) + k_y(t))\cos\omega_0 t \end{pmatrix} \tag{6.127}$$

This is the memory kernel in the non-Markovian eq. (6.122). As in the last section, this equation is correct as it stands (when V is small), but because R can be fast, $\langle \delta\rho; t \rangle$ is not necessarily slow, and it is not safe to make a Markovian approximation. First we must remove R, then make the Markovian approximation, and finally restore R. So we change variables,

$$\langle \delta\sigma; t \rangle = e^{tR} \cdot g(t). \tag{6.128}$$

Then g obeys

$$\frac{\partial}{\partial t} g(t) = \int_0^t dt' e^{-tR} \cdot K(t-t') \cdot e^{tR} \cdot g(t')$$

$$= \int_0^t dt' e^{-tR} \cdot K(t') \cdot e^{(t-t')R} \cdot g(t-t'). \tag{6.129}$$

If the fluctuations are small, K is small, and g is a slow variable. Then we can replace $g(t-t')$ by $g(t)$ and extend the t' integral to infinity,

$$\frac{\partial}{\partial t} g(t) \cong e^{-tR} \cdot \left(\int_0^\infty dt' K(t') \cdot e^{-t'R} \right) \cdot e^{tR} \cdot g(t). \tag{6.130}$$

Restoring R, we get the correct Markovian approximation to eq. (6.122),

$$\frac{\partial}{\partial t}\langle\delta\sigma;t\rangle \cong R\cdot\langle\delta\sigma;t\rangle + \left(\int_0^\infty dt' K(t')\cdot e^{-t'R}\right)\cdot\langle\delta\sigma;t\rangle. \quad (6.131)$$

When the matrix product is worked out, the resulting Bloch equations contain a new frequency ω_1 and two relaxation times, T_1 and T_2,

$$\frac{\partial}{\partial t}\langle\sigma_x\rangle = \omega_1\langle\sigma_y\rangle - \frac{1}{T_2}\langle\sigma_x\rangle \quad (6.132)$$

$$\frac{\partial}{\partial t}\langle\sigma_y\rangle = -\omega_1\langle\sigma_x\rangle - \frac{1}{T_2}\langle\sigma_y\rangle \quad (6.133)$$

$$\frac{\partial}{\partial t}\langle\sigma_z\rangle = -\frac{1}{T_1}(\langle\sigma_z\rangle - \langle\sigma_z\rangle_{eq}). \quad (6.134)$$

(Here we assume that $k_x = k_y$.) The constants are

$$\omega_1 = \omega_0 - \int_0^\infty dt \sin\omega_0 t\, k_x(t) \quad (6.135)$$

$$\frac{1}{T_1} = 2\int_0^\infty dt \cos\omega_0 t\, k_x(t) \quad (6.136)$$

$$\frac{1}{T_2} = \int_0^\infty dt(k_z(t) + \cos\omega_0 t\, k_x(t)). \quad (6.137)$$

The first two Bloch equations can be combined to give

$$\frac{\partial}{\partial t}\langle\sigma_x + i\sigma_y\rangle = -\left(i\omega_1 + \frac{1}{T_2}\right)\langle\sigma_x + i\sigma_y\rangle \quad (6.138)$$

and its complex conjugate. These can be solved easily for the averages and used to find the time correlation function, for example, of σ_x,

$$\langle\sigma_x(t)\sigma_x(0)\rangle = \cos(\omega_1 t)e^{-t/T_2}. \quad (6.139)$$

Suppose the nucleus is subjected to a periodic external magnetic field in the x direction, so the perturbation Hamiltonian is

$$\mathcal{H}_{ext} = V_{ext}\cos(\omega t)\sigma_x. \quad (6.140)$$

The energy absorption, as in section 3.2, is determined by

$$\int_0^\infty dt \cos(\omega t)\langle\sigma_x(t)\sigma_x(0)\rangle = \int_0^\infty dt\, e^{-i\omega t}\cos(\omega_1 t)e^{-t/T_2} \quad (6.141)$$

and is therefore proportional to

QUANTUM DYNAMICS   121

$$\left(\frac{1}{1+(\omega-\omega_1)^2 T_2^2}+\frac{1}{1+(\omega+\omega_1)^2 T_2^2}\right). \tag{6.142}$$

The absorption spectrum consists of Lorentzians centered at frequencies $\pm\omega_1$, each with the width $1/T_2$. Note that there are two distinct contributions to this line width, one from k_x and one from k_z.

This argument rested on the local equilibrium approximation in eq. (6.121). This can be justified in exactly the same way as in the preceding chapter. The error made in this approximation contributes to the rate of change only in the third order of V and cannot affect the second-order result.

6.5 Master Equation Revisited

The quantum mechanical or Pauli master equation, discussed extensively in section 3.3, treats the effects of a weak perturbation on transitions between unperturbed quantum states. Here the general method used to treat two-state models in the preceding chapters is applied to deriving the master equation. After that, a projection operator method is used to derive a generalized master equation that is formally valid to all orders of the perturbation.

As in section 3.3, the total Hamiltonian H consists of an unperturbed energy H_0 and an interaction energy V giving rise to transitions between unperturbed states,

$$H = H_0 + V. \tag{6.143}$$

In the unperturbed representation, H_0 is diagonal and V has no diagonal elements. The probability of finding the system in the jth unperturbed state at time t (or the diagonal part of the density matrix ρ in the unperturbed representation) is $\rho_{jj}(t)$.

The derivation starts with the quantum mechanical Liouville equation for the time-dependent density matrix $\rho(t)$,

$$\frac{d}{dt}\rho(t) = -(L_0 + L_V)\rho(t), \tag{6.144}$$

where $L = L_0 + L_V$ is the Liouville operator. Because the unperturbed Hamiltonian is diagonal, the tetradic form of the operators are

$$(L_0)_{jk,lm} = \frac{i}{\hbar}(E_j - E_k)\delta_{jl}\delta_{km} \tag{6.145}$$

$$(L_V)_{jk,lm} = \frac{i}{\hbar}(V_{jl}\delta_{km} - \delta_{jl}V_{mk}), \tag{6.146}$$

and the tetradic form of the Liouville equation is

$$\frac{d}{dt}\rho_{jk}(t) = -\sum_{l,m} L_{jk,lm} \rho_{lm}(t). \tag{6.147}$$

Following the same scheme as in the last two sections, we integrate the Liouville equation once,

$$\rho(t) = e^{-L_0 t}\rho(0) - \int_0^t ds\, e^{-L_0(t-s)} L_V \rho(s), \tag{6.148}$$

and substitute this back into the Liouville equation,

$$\frac{\partial}{\partial t}\rho(t) = -L_0\rho(t) - L_V e^{-L_0 t}\rho(0) + L_V \int_0^t ds\, e^{-L_0(t-s)} L_V \rho(s). \tag{6.149}$$

This is still exact.

The master equation contains only the diagonal elements of the density matrix, but the Liouville equation mixes diagonal and off-diagonal elements. On taking the diagonal part, the first term in the right hand side drops out, $(L_0\rho)_{jj} = 0$, because the unperturbed Hamiltonian is diagonal. We restrict our attention to only those situations where the density matrix is initially diagonal (often called the assumption of initial random phase),

$$\rho_{jk}(0) = \rho_{jj}(0)\delta_{jk}. \tag{6.150}$$

Then, $\exp(-L_0 t)\rho(0)$ remains diagonal and the second term in the right hand side drops out because $(L_V)_{jj,ll} = 0$. The third part of the right hand side remains,

$$\frac{d}{dt}\rho_{jj}(t) = \sum_{kl}\left(L_V \int_0^t ds\, e^{-L_0(t-s)} L_V\right)_{jj,kl} \rho_{kl}(s). \tag{6.151}$$

If the initial density matrix is diagonal, this is exact.

Now we make the same crucial approximation as in the preceding sections; we replace the complete density matrix in the sum by its diagonal part,

$$\rho_{kl}(s) \to \rho_{kk}(s)\delta_{kl}. \tag{6.152}$$

The result (after switching t and $t-s$) is

$$\frac{d}{dt}\rho_{jj}(t) \cong \sum_k \int_0^t ds (L_V e^{-L_0 s} L_V)_{jj,kk} \rho_{kk}(t-s). \tag{6.153}$$

Now we can write

$$(L_V e^{-Ls} L_V)_{jj,kk} = K_{jj,kk}(s) \tag{6.154}$$

and

$$\frac{d}{dt}\rho_{jj}(t) \cong \int_0^t ds \sum_k K_{jj,kk}(s)\rho_{kk}(t-s). \tag{6.155}$$

The memory kernels satisfy a sum rule,

$$\sum_k K_{jj,kk}(s) = 0, \tag{6.156}$$

which is a direct consequence of

$$\sum_k (L_V)_{ij,kk} = 0. \tag{6.157}$$

On using the sum rule, eq. (6.155) becomes

$$\frac{d}{dt}\rho_{jj}(t) \cong \int_0^t ds \sum_{k \ne j} K_{jj,kk}(s)(\rho_{kk}(t-s) - \rho_{jj}(t-s)). \tag{6.158}$$

We need to evaluate $K_{jj,kk}$ only for $j \ne k$. The memory kernel becomes

$$K_{jj,kk} = \sum_{ab}\sum_{cd}(L_V)_{jj,ab}(e^{-sL_0})_{ab,cd}(L_V)_{cd,kk}$$

$$= \sum_{ab}\sum_{cd}\left(\frac{i}{\hbar}\right)^2 (V_{ja}\delta_{jb} - V_{bj}\delta_{ja})e^{-isE_a/\hbar}\delta_{ac}e^{isE_b/\hbar}\delta_{bd}(V_{ck}\delta_{dk} - V_{kd}\delta_{ck}). \tag{6.159}$$

When the sums and products are worked out, one gets (for $j \ne k$)

$$K_{jj,kk}(s) = \frac{1}{\hbar^2}|V_{jk}|^2 2\cos(\omega_{jk}s), \tag{6.160}$$

where $\omega_{jk} = (E_j - E_k)/\hbar$. The resulting equation is

$$\frac{d}{dt}\rho_{jj}(t) \cong \int_0^t ds \sum_{k \ne j} \frac{2}{\hbar^2}|V_{jk}|^2 \cos\omega_{jk}s(\rho_{kk}(t-s) - \rho_{jj}(t-s)). \tag{6.161}$$

The rate of change of any diagonal element is of the order of V^2, and so it is a good approximation to replace $\rho_{jj}(t-s)$ and $\rho_{kk}(t-s)$ by $\rho_{jj}(t)$ and $\rho_{kk}(t)$ in the right hand side (the Markovian approximation). Then the time integral over s can be done, and the result is

$$\frac{d}{dt}\rho_{jj}(t) \cong \sum_{k \ne j} \frac{2}{\hbar^2}|V_{jk}|^2 \frac{\sin\omega_{jk}t}{\omega_{jk}}(\rho_{kk}(t) - \rho_{jj}(t)). \tag{6.162}$$

For large t, the quantity $\sin(\omega t)/\omega$ is very much like a delta function in ω; its height is t and its width is of order $1/t$, and its integral over ω is π. If the delta function approximation is inserted in eq. (6.162), one gets the standard Pauli master equation,

$$\frac{d}{dt}\rho_{jj}(t) \cong \sum_{k \neq j} \frac{2\pi}{\hbar}|V_{jk}|^2 \delta(\hbar\omega_{jk})(\rho_{kk}(t) - \rho_{jj}(t)), \tag{6.163}$$

containing transition rates given by the Golden Rule.

For this to be correct, one needs to find many unperturbed states within the width of the function $\sin(\omega t)/\omega$ that led to the delta function. This width is small, of the order of $1/t$ because t is large, of the order of $1/V^2$. Fortunately, in typical applications of the Golden Rule, the unperturbed states have energy levels that are inverse to the size of the system, either a large number of particles n or a large volume v. Then the spectrum is almost dense, and the delta function approximation is useful as long as $t \ll O(n)$ or $O(v)$.

Aside from the choice of a special initial state, the only approximation that was made is eq. (6.152). But a simple calculation, like the one done in the previous chapters, leads to $\rho_{kl}(s) - \rho_{kk}(s)\delta_{kl} = O(V\rho)$. The error is therefore of third order in V.

Projection Operator Method

As in the preceding sections, at a particular point of the derivation an exact distribution function was replaced by a simpler one that carried only limited information. In the present case, this can be done by means of a projection operator (S. Nakajima, 1958; R. Zwanzig, 1960). This is a simple illustration of a general procedure that will be treated more completely in following chapters. The diagonal part of any matrix A is obtained by the tetradic operator P,

$$P_{jk,lm} = \delta_{jk}\delta_{jl}\delta_{km}, \tag{6.164}$$

so that

$$(PA)_{jk} = \sum_{kl} P_{jk,lm} A_{lm} = \delta_{jk} A_{jj}. \tag{6.165}$$

By repeating the process, one sees immediately that $PPA = PA$. The property $P^2 = P$ means that P is a projection operator. This operator will be used to separate diagonal and off-diagonal parts of the density matrix.

The master equation involves only the diagonal elements of the density matrix, but the Liouville equation mixes diagonal and off-diagonal elements. When $j = k$, L_0 drops out, and the Liouville equation becomes

$$\frac{\partial}{\partial t}\rho_{jj} = -\sum_{lm}(L_V)_{jj,lm}\rho_{lm}. \tag{6.166}$$

Because V has no diagonal part, only the off-diagonal part of the density matrix contributes to the right hand side. This suggests that we should find out how the off-diagonal part is controlled by the diagonal part.

To do this, the entire density matrix is partitioned into diagonal and off-diagonal parts by projecting with P,

$$\rho = \rho^{(d)} + \rho^{(od)}, \qquad \rho^{(d)} = P\rho, \qquad \rho^{(od)} = (1-P)\rho, \tag{6.167}$$

and

$$[\rho^{(d)}]_{jk} = \rho_{jj}\delta_{jk}, \qquad [\rho^{(od)}]_{jk} = \rho_{jk}(1-\delta_{jk}). \tag{6.168}$$

Operating on the Liouville equation by P and $(1-P)$, one gets two coupled equations,

$$\frac{d}{dt}\rho^{(d)} = -PL\rho^{(d)} - PL\rho^{(od)} \tag{6.169}$$

$$\frac{d}{dt}\rho^{(od)} = -(1-P)L\rho^{(d)} - (1-P)L\rho^{(od)}. \tag{6.170}$$

We have already looked at the first of these equations; because $PL_0 = 0$, only the second term survives and L is replaced by L_V. Earlier we found $\rho^{(od)}$ by an iterative procedure. Now we solve for it formally. The solution uses the exponential operator $\exp((1-P)Lt)$. It contains the initial value of $\rho^{(od)}$ and the history of $\rho^{(d)}$ from 0 to t,

$$\rho^{(od)}(t) = -\int_0^t ds\, e^{-(1-P)L(t-s)}(1-P)L\rho^{(d)}(s) + e^{-(1-P)Lt}\rho^{(od)}(0). \tag{6.171}$$

This is substituted into the first equation, leading to

$$\frac{d}{dt}\rho^{(d)}(t) = PL\int_0^t ds\, e^{-(1-P)L(t-s)}(1-P)L\rho^{(d)}(s) - PLe^{-(1-P)Lt}\rho^{(od)}(0). \tag{6.172}$$

As before, we limit ourselves to situations where the density matrix is initially diagonal. Then $\rho^{(od)}(0) = 0$, and the second term in this equation can be dropped. The result is a formally exact equation that contains only the diagonal part of the density matrix. Because $PL_0 = 0$ and $L_0P = 0$, the initial and final Liouville operators in the last equation can be replaced by L_V. On switching t and $t-s$, eq. (6.172) becomes

$$\frac{d}{dt}\rho_{jj}(t) = \sum_k \int_0^t ds (L_V e^{-(1-P)Ls}(1-P)L_v)_{jj,kk} \rho_{kk}(t-s)$$

$$= \sum_k \int_0^t ds K_{jj,kk}(s)\rho_{kk}(t-s). \tag{6.173}$$

The memory kernel here is a generalization of the one found earlier:

$$K_{jj,kk}(s) = (L_V e^{-(1-P)(L_0+L_V)s}(1-P)L_V)_{jj,kk}. \tag{6.174}$$

This is formally second order in V, and more V dependence comes from the Liouville operator in the exponent. These memory kernels satisfy the same sum rule as before.

While eq. (6.173) is formally exact, it is not particularly useful because of the projection operator in the exponent. If, however, we are only interested in terms of order V^2, L_V can be dropped from the exponent:

$$K_{jj,kk}(s) \to (L_V e^{-(1-P)L_0 s}(1-P)L_V)_{jj,kk} + O(V^3). \tag{6.175}$$

Then the projection operator in the exponent can be removed because the tetradic product $PL_0 = 0$. The memory kernel becomes

$$K_{jj,kk}(s) = (L_V e^{-sL_0} L_V)_{jj,kk} + O(V^3). \tag{6.176}$$

This is precisely the expression found earlier.

Heat Bath Master Equation

The same procedure can be used to derive the heat bath master equation discussed in section 3.3. The projection operator has two parts. First, the actual bath dependence is integrated out and replaced by the equilibrium bath distribution. Second, the system density matrix is replaced by its diagonal part. With an initial condition of this form (diagonal in system, equilibrium in bath), the derivation follows easily.

7

Linear Response Theory

7.1 Static Linear Response

A system is in a state of thermal equilibrium, and then a weak external field is turned on. How does the system respond? For example, we might want to know how much current is induced in an ionic solution by an electric field. This is the kind of question treated by linear response theory.

If the applied field is held constant for a very long time, so that the system can come to equilibrium in the presence of the field, finding the response is a problem of equilibrium statistical mechanics. But if we want to know the transient response to the applied field, or if the field varies periodically in time, then it is necessary to go beyond equilibrium statistical mechanics. This section deals with the response to a static force. The following section presents the theory for a time-dependent force.

It is useful to see first how the equilibrium linear response is determined. The system is described by an unperturbed Hamiltonian $H(X)$. The applied field is denoted by E. The coupling of the system to the field is described by the energy $-M(X)E$, where $M(X)$ is some known function of the state of the system. For the following discussion, the exact nature of E and M is not important; but it is helpful to keep in mind the often-used example where E is an electric field and M is a total electric dipole moment. The perturbed Hamiltonian is $H(X, E) = H(X) - M(X)E$.

First we calculate the response using classical mechanics. The unperturbed distribution function denoted $f(X)$ is

$$f(X) = \frac{1}{Q} e^{-\beta H(X)}, \qquad Q = \int dX e^{-\beta H(X)}, \tag{7.1}$$

where Q is the unperturbed partition function. Averages taken with the unperturbed distribution are denoted by $\langle\ \rangle$. The corresponding perturbed quantities are

$$f(X; E) = \frac{1}{Q(E)} e^{-\beta H(X) + \beta M(X)E} \tag{7.2}$$

$$Q(E) = \int dX e^{-\beta H(X) + \beta M(X)E}. \tag{7.3}$$

In classical statistical mechanics, one can easily expand the perturbed system about the unperturbed system. This is where the corresponding quantum mechanical treatment differs; if the operators H and M do not commute, one has to be careful about ordering operators. We return to this later. To first order, that is, in linear response, one finds

$$e^{-\beta H + \beta ME} = \{1 + \beta ME + O(E^2)\} e^{-\beta H} \tag{7.4}$$

$$Q(E) = Q\{1 + \beta \langle M \rangle E + O(E^2)\}, \tag{7.5}$$

so that the expansion of the perturbed distribution is

$$f(X; E) = \{1 + \beta[M(X) - \langle M \rangle]E\} f(X) + O(E^2). \tag{7.6}$$

These equations contain the unperturbed equilibrium average $\langle M \rangle$, coming from the first-order expansion of $Q(E)$. For simplicity, from here on we restrict the discussion to cases where $\langle M \rangle = 0$.

At this point, we can ask for the average of any dynamical variable $A(X)$, denoted by $\langle A; E \rangle$. $A(X)$ could be, for example, the quantity $M(X)$ itself. The result is

$$\langle A; E \rangle = \langle A \rangle + \chi_{AM} E + O(E^2), \tag{7.7}$$

where the coefficient χ_{AM} is given by

$$\chi_{AM} = \beta \langle AM \rangle. \tag{7.8}$$

If we choose for A the total electric dipole moment M, then χ_{AM} is a kind of dielectric susceptibility. Whatever our choice of A, the quantity

χ_{AM} describes the average linear response $\langle A; E \rangle$ produced by the applied field.

The quantum mechanical version of this derivation uses an operator expansion of the perturbed distribution function. This expansion can be derived, for example, by starting with the Laplace transform of the equilibrium distribution with respect to β,

$$\int_0^\infty d\beta\, e^{-\beta(H-ME)} e^{-z\beta} = \frac{1}{z+H-ME}. \tag{7.9}$$

This quantity obeys the operator identity

$$\frac{1}{z+H-ME} = \frac{1}{z+H} + \frac{1}{z+H} ME \frac{1}{z+H-ME}, \tag{7.10}$$

or, to first order in E,

$$\frac{1}{z+H-ME} = \frac{1}{z+H} + \frac{1}{z+H} ME \frac{1}{z+H} + O(E^2). \tag{7.11}$$

On inverting the Laplace transform, the second term leads to a convolution,

$$e^{-\beta H + \beta ME} = e^{-\beta H} + \int_0^\beta d\lambda\, e^{-\lambda H} ME e^{-(\beta-\lambda)H} + O(E^2). \tag{7.12}$$

(Note that since this is a convolution, one can switch λ and $\beta - \lambda$, thereby obtaining expressions that are equivalent but look different.) For notational convenience, we take advantage of the Heisenberg representation of the time dependence of any dynamical quantity, using the imaginary time $i\hbar\lambda$,

$$e^{-\lambda H} M e^{\lambda H} = M(i\hbar\lambda), \tag{7.13}$$

so that the expansion becomes

$$e^{-\beta(H-ME)} = e^{-\beta H} + \int_0^\beta d\lambda\, M(i\hbar\lambda) E e^{-\beta H} + \cdots. \tag{7.14}$$

One further bit of notation is helpful. We define the "Kubo transform" of the operator M by a tilde,

$$\tilde{M} = \frac{1}{\beta} \int_0^\beta d\lambda\, M(i\hbar\lambda). \tag{7.15}$$

130   NONEQUILIBRIUM STATISTICAL MECHANICS

In the classical limit, this approaches M. (Note that if the equilibrium average of M vanishes, so does the average of its Kubo transform.) The partition function has the expansion

$$Q(E) = Q + O(E^2) \cdots, \qquad (7.16)$$

and the susceptibility is

$$\chi_{AM} = \beta \langle A\tilde{M} \rangle. \qquad (7.17)$$

The quantum perturbation theory differs from the classical theory only in the replacement of M by its Kubo transform.

An identity derived by switching $\beta - \lambda$ and λ is

$$\langle A\tilde{M} \rangle = \langle M\tilde{A} \rangle. \qquad (7.18)$$

7.2 Dynamic Linear Response

Linear Response in Classical Mechanics

The dynamical version of the preceding perturbation theory is quite straightforward. To find the time-dependent average of a dynamical variable A, we use the time-dependent distribution function $f(X; t)$, which evolves from some given initial distribution function $f(X; 0)$. As in the equilibrium theory, we look for the way that $f(X; t)$ is affected by the extra perturbing Hamiltonian $-M(X)E(t)$, where $E(t)$ is now a time-dependent external field.

The time-dependent distribution function obeys the Liouville equation,

$$\frac{\partial f}{\partial t} = -L_0 f - L_1 E(t) f, \qquad (7.19)$$

in which L_0 is the unperturbed Liouville operator. $L_0 f$ is the Poisson bracket of H and f, and $L_1 f$ is the Poisson bracket of $-M$ and f. To find the first-order response to $E(t)$, we expand f in powers of E, using f_0 and f_1 to denote the terms of zeroth and first order in E,

$$f = f_0 + f_1 + O(E^2). \qquad (7.20)$$

Then on collecting terms of order zero and one, the Liouville equation gives

$$\frac{\partial f_0}{\partial t} = -L_0 f_0 \qquad (7.21)$$

LINEAR RESPONSE THEORY   131

$$\frac{\partial f_1}{\partial t} = -L_0 f_1 - L_1 E(t) f_0. \tag{7.22}$$

Suppose that the system is in thermal equilibrium before the field is turned on (at $t = 0$). This initial condition is the one most commonly used in applications and is the only one considered here. Then $f(0) = f_{eq}$, and the preceding two equations are to be solved with the initial conditions,

$$f_0(0) = f_{eq}, \qquad f_1(0) = 0. \tag{7.23}$$

The first equation has the obvious solution

$$f_0(t) = f_{eq} \quad \text{(all } t\text{)} \tag{7.24}$$

because $L_0 f_{eq} = 0$. The second equation is an inhomogeneous first-order differential equation, and the initial value vanishes, so it has the operator solution

$$f_1(t) = -\int_0^t ds\, e^{-(t-s)L_0} L_1 E(s) f_0(s). \tag{7.25}$$

But $f_0(s)$ may be replaced by f_{eq}. Recall that $L_1 f_{eq}$ is a Poisson bracket, and written out in a more explicit form,

$$L_1 f_{eq} = -\left[\frac{\partial M}{\partial \mathbf{p}} \cdot \frac{\partial f_{eq}}{\partial \mathbf{q}} - \frac{\partial M}{\partial \mathbf{q}} \cdot \frac{\partial f_{eq}}{\partial \mathbf{p}}\right]. \tag{7.26}$$

But because f_{eq} is $\exp(-\beta H)/Q$, the derivatives transform to

$$= +\beta f_{eq}\left[\frac{\partial M}{\partial \mathbf{p}} \cdot \frac{\partial H}{\partial \mathbf{q}} - \frac{\partial M}{\partial \mathbf{q}} \cdot \frac{\partial H}{\partial \mathbf{p}}\right]. \tag{7.27}$$

The square bracket in this equation is in fact minus the Poisson bracket of H and M, or $L_0 M$, or the time derivative (overdot) of M,

$$L_1 E(s) f_{eq} = -\beta E(s) \dot{M} f_{eq}. \tag{7.28}$$

Then the perturbation to f is

$$f_1(t) = \int_0^t ds\, \beta E(s) e^{-(t-s)L_0} \dot{M} f_{eq}. \tag{7.29}$$

Now we use this to find the time-dependent average of some quantity $A(X)$, The result is

132   NONEQUILIBRIUM STATISTICAL MECHANICS

$$\langle A;t\rangle = \langle A\rangle_{eq} + \beta\int_0^t ds E(s)\int dX A(X)e^{-(t-s)L_0}\dot{M}f_{eq}. \quad (7.30)$$

In typical applications, the equilibrium averages of both A and M vanish. From here on we assume that this is so. The exponential Liouville operator can work backwards on $A(X)$, generating the time dependent $A(t-s; X)$; and the phase space integral gives an equilibrium average,

$$\langle A;t\rangle = \beta\int_0^t ds E(s)\langle A(t-s)\dot{M}(0)\rangle_{eq} + \cdots. \quad (7.31)$$

This suggests defining the time-dependent analog of the static susceptibility,

$$\phi_{AM}(t) = \beta\langle A(t)\dot{M}(0)\rangle_{eq}. \quad (7.32)$$

Then, after switching $t-s$ and s, we have obtained the standard linear response formula (R. Kubo, 1957),

$$\langle A;t\rangle = \beta\int_0^t ds\,\phi_{AM}(s)E(t-s) + O(E^2). \quad (7.33)$$

If a constant external field is switched on at $t=0$, and we ask for the response at infinite time, the preceding equation becomes

$$\langle A;\infty\rangle = \chi_{AM}E, \qquad \chi_{AM} = \int_0^\infty dt\,\phi_{AM}(t). \quad (7.34)$$

To verify this integral representation of χ_{AM}, note that in ϕ, the time derivative can be moved from M to $A(t)$, with a change of sign. The time integral of the time derivative of A is $A(\infty) - A(0)$. The equilibrium average $\langle A(\infty)M(0)\rangle$ vanishes; at infinite time, there is no correlation of $A(t)$ and $M(0)$. What remains is just $\beta\langle A(0)M(0)\rangle_{eq} = \chi_{AM}$.

Because the preceding calculation was restricted to first order in the perturbation, the effects of different external forces $E_j(t)$ are additive. Further, one may ask for the time dependence of a number of different response functions A_i. If the perturbation has the form $-\Sigma_i M_i E_i(t)$, then the total response is a sum of the responses to individual forces $E_i(t)$, as

$$\langle A_i;t\rangle = \sum_j \int_0^t ds\,\phi_{ij}(t-s)E_j(s) + \cdots \quad (7.35)$$

$$\phi_{ij}(t) = \beta\langle A_i(t)\dot{M}_j(0)\rangle_{eq}. \quad (7.36)$$

Linear Response in Quantum Mechanics

The quantum mechanical version of this theory differs from the classical mechanical theory in three important ways. First, the phase space average of a dynamical variable A is replaced by the quantum average,

$$\langle A \rangle_{\text{classical}} = \int d\mathbf{X}\, A(\mathbf{X}) f(\mathbf{X}, t) \to \langle A \rangle_{\text{quantum}} = \text{Trace } A \cdot \rho(t). \tag{7.37}$$

Second, the classical Liouville operator, defined as a Poisson bracket, is replaced by the quantum commutator,

$$L_{\text{classical}} = \{H,\ \}_{\text{Poisson bracket}} \to L_{\text{quantum}} = \frac{i}{\hbar}[H,\ ]_{\text{commutator}}. \tag{7.38}$$

Third, the classical Liouville equation for the phase space distribution function is replaced by the quantum Liouville equation for the density matrix,

$$\frac{\partial f}{\partial t} = -L_{\text{classical}} f \to \frac{\partial \rho}{\partial t} = -L_{\text{quantum}} \rho. \tag{7.39}$$

Despite these changes, most of the preceding derivation carries over without change in quantum mechanics.

The Hamiltonian is still $H(t) = H_0 + H_1 E(t)$. The corresponding quantum Liouville operator is still $L(t) = L_0 + L_1 E(t)$. The quantum Liouville equation is formally the same,

$$\frac{\partial \rho}{\partial t} = -L_0 \rho - L_1 E(t) \rho. \tag{7.40}$$

We expand ρ to first order in E,

$$\rho(t) = \rho_0(t) + \rho_1(t) + O(E^2) \tag{7.41}$$

and get equations for the two parts,

$$\frac{\partial \rho_0}{\partial t} = -L_0 \rho_0, \qquad \rho_0(t=0) = \rho_{\text{eq}} \tag{7.42}$$

and

$$\frac{\partial \rho_1}{\partial t} = -L_0 \rho_1 - L_1 E(t) \rho_0, \qquad \rho_1(t=0) = 0. \tag{7.43}$$

As before, the first equation has the solution,

$$\rho_0(t) = \rho_{eq} \quad \text{(all } t\text{)}, \tag{7.44}$$

and the formal operator solution of the second equation is

$$\rho_1(t) = -\int_0^t ds\, e^{-(t-s)L_0} L_1 E(s) \rho_{eq}. \tag{7.45}$$

The average of some variable A is

$$\langle A; t \rangle = \int_0^t ds\, \phi(t-s) E(s) + \cdots, \tag{7.46}$$

where the response function is

$$\phi(t) = -\text{Trace } A e^{-tL_0} L_1 \rho_{eq} = -\text{Trace } A(t) L_1 \rho_{eq}. \tag{7.47}$$

Now we encounter another version of the difficulty seen already in the quantum theory of the static susceptibility. In the classical mechanical theory, we were able to derive $L_1 f_{eq} = \beta f_{eq} L_0 H_1$, but the derivation fails in quantum mechanics because the operators do not commute. There are several ways around this difficulty. The first uses the definition of the Liouville operator as a commutator and the invariance of a trace to a cyclic permutation,

$$\phi(t) = -\text{Trace } A(t) \frac{i}{\hbar}[H_1, \rho_{eq}] = -\text{Trace } \frac{i}{\hbar}[A(t), H_1]\rho_{eq}$$
$$= \langle L_1 A(t) \rangle_{eq}. \tag{7.48}$$

This does not look much like a conventional time correlation function. However, as we shall see shortly, there are many ways of writing $\phi(t)$. Another approach starts with the commutator of H_1 and $\exp(-\beta H_0)$, and with the definition

$$\Phi = [H_1, e^{-\beta H_0}]e^{\beta H_0} = H_1 - e^{-\beta H_0} H_1 e^{\beta H_0}. \tag{7.49}$$

This quantity vanishes at $\beta = 0$ and satisfies the differential equation

$$\frac{\partial}{\partial \beta}\Phi(\beta) = e^{-\beta H_0}[H_0, H_1]e^{\beta H_0}. \tag{7.50}$$

On integrating over β, we obtain

$$\Phi(\beta) = \int_0^\beta d\lambda\, e^{-\lambda H_0}[H_0, H_1]e^{\lambda H_0}, \tag{7.51}$$

which leads to

$$L_1 \rho_{eq} = \int_0^\beta d\lambda \, e^{-\lambda H_0} (L_0 H_1) e^{\lambda H_0} \rho_{eq}. \tag{7.52}$$

One should not be surprised that this contains the Kubo transform, because it turned up in the earlier treatment of the static quantum susceptibility; then,

$$L_1 \rho_{eq} = \beta L_0 \tilde{H}_1 \rho_{eq}. \tag{7.53}$$

The response function involves the time correlation function of the dynamical variable A at the real time t, and the perturbation $\tilde{H}_1$,

$$\phi(t) = -\beta \langle A(t) L_0 \tilde{H}_1 \rangle_{eq}. \tag{7.54}$$

As in the classical case, this may be written in a variety of forms by moving around Liouville operators and Kubo transforms.

Frequency Dependent Response

The frequency-dependent form of the preceding linear response formula is of great interest because of its utility in describing certain experiments. Quite often, a measurement of some linear response is made by switching on a periodically varying field, waiting until transients have died out, and then measuring the response at the frequency of the perturbation. For example, if the perturbation is a periodic electric field and the response is an induced electric current, the results of the measurement are described by a frequency-dependent conductivity.

The entire measurement process is contained in eq. (7.33). An arbitrary field is switched on at $t = 0$; this means that $E(t) = 0$ for earlier times $t < 0$. The response $\langle A; t \rangle$ is obtained for later times $t > 0$ and vanishes for earlier times $t < 0$. (This is "causality"—a response cannot come before the cause of the response.) The Fourier transforms of these quantities are

$$E_\omega = \int_{-\infty}^\infty dt \, e^{-i\omega t} E(t) = \int_0^\infty dt \, e^{-i\omega t} E(t) \tag{7.55}$$

$$\langle A \rangle_\omega = \int_{-\infty}^\infty dt \, e^{-i\omega t} \langle A; t \rangle = \int_0^\infty dt \, e^{-i\omega t} \langle A; t \rangle. \tag{7.56}$$

The second form of each transform is allowed because in this experiment both E and $\langle A \rangle$ vanish for negative t. In fact these integrals are Laplace transforms, with the transform variable z replaced by $i\omega$. Now take the Fourier transform of eq. (7.33). This is also the Laplace transform of that equation, and we know that the Laplace transform of a convolution is the product of the individual transforms,

$$\int_0^\infty dt \exp(-i\omega t)\int_0^t ds\,\phi(s)E(t-s)$$
$$=\int_0^\infty ds\exp(-i\omega s)\phi(s)\times\int_0^\infty dt\exp(-i\omega t)E(t). \qquad (7.57)$$

Then eq. (7.33) becomes

$$\langle A\rangle_\omega = \sigma_{AM}(\omega)E_\omega \qquad (7.58)$$

$$\sigma_{AM}(\omega) = \int_0^\infty dt\,e^{i\omega t}\phi_{AM}(t). \qquad (7.59)$$

This equation, containing the frequency-dependent response function $\sigma(\omega)$, is valid for all frequencies, no matter what the actual time dependence of E was. If a field with frequency ω_0 is switched on at $t = 0$, its Fourier transform contains components at all other frequencies, associated with the switching-on step function; and the transform of the response also contains components at all other frequencies, describing the decay of initial transients.

The response function $\sigma(\omega)$ is a one-sided Fourier transform of $\phi(t)$ and is complex. Its real part is the cosine transform of $\phi(t)$, and its imaginary part is the sine transform. Suppose that an experiment provides only the frequency dependence of the imaginary part of $\sigma(\omega)$: Then, by inverting the sine transform, the time dependence of $\phi(t)$ can be found. Taking the cosine transform provides the real part of $\sigma(\omega)$. Thus the imaginary part determines the real part. The procedure works backwards too; the real part determines the imaginary part. This is called the Kramers-Kronig relation. To use it in practice, one needs good data over a wide range of frequencies.

7.3 Applications of Linear Response Theory

This section presents several applications of linear response theory. The first application is to the calculation of the mobility of a single ion in solution. The ion has a charge e and interacts with an external uniform electric field E in the x direction. The perturbation Hamiltonian is $-eE(t)x$. (This is the product of a charge e and the electrostatic potential $-E(t)x$.) The quantity M is $e\,x$. The mobility of the ion is its average velocity $\langle v\rangle$, so we choose

$$A = \dot{x} = v. \qquad (7.60)$$

Then linear response theory leads to a formula for the mobility of the ion,

LINEAR RESPONSE THEORY   137

$$\langle v \rangle_\omega = \mu(\omega) E_\omega \qquad (7.61)$$

$$\mu(\omega) = \int_0^\infty dt\, e^{-i\omega t} \beta \langle v(t)v(0) \rangle. \qquad (7.62)$$

The mobility is determined by the ion's velocity correlation function. (This quantity was introduced in section 1.2; its time integral is a diffusion coefficient.) If the velocity correlation function decays exponentially, as in Brownian motion theory,

$$\langle v(t)v(0) \rangle_{eq} = \frac{kT}{m} \exp(-\zeta t/m), \qquad (7.63)$$

then the frequency-dependent mobility is

$$\sigma(\omega) = \frac{e}{i\omega m + \zeta}. \qquad (7.64)$$

The zero frequency limit is e/ζ.

Another standard application of linear response theory is to find an expression for the frequency-dependent magnetic susceptibility of a material. In this case, M is the total magnetic moment of the system, $E(t)$ is replaced by the time-dependent magnetic field $B(t)$, and A is taken to be the magnetization M/V, where V is the volume of the system. The magnetic susceptibility χ is defined by

$$\frac{1}{V} \langle M \rangle_\omega = \chi(\omega) B_\omega. \qquad (7.65)$$

Then linear response theory provides

$$\chi(\omega) = \frac{\beta}{V} \int_0^\infty dt\, e^{-i\omega t} \langle M(t) L_0 \tilde{M} \rangle. \qquad (7.66)$$

A rearrangement, using the adjoint properties of L_0, leads to a more useful form. As in the earlier general discussion, the Liouville operator can act to the left on $M(t)$, producing the time derivative $-dM(t)/dt$, and the time derivative can be moved in front of the average,

$$\chi(\omega) = \frac{\beta}{V} \int_0^\infty dt\, \exp(-i\omega t) \left(-\frac{d}{dt}\right) \langle M(t) \tilde{M} \rangle_{eq}. \qquad (7.67)$$

Next, an integration by parts is performed:

$$\int_0^\infty dt\, e^{-i\omega t}\frac{d}{dt}\langle M(t)\tilde{M}\rangle_{eq}$$
$$=\lim_{t\to\infty} e^{-i\omega t}(\langle M(t)\tilde{M}\rangle_{eq} - \langle M(0)\tilde{M}\rangle_{eq}) + i\omega\int_0^\infty dt\, e^{-i\omega t}\langle M(t)\tilde{M}\rangle_{eq}. \quad (7.68)$$

At long times, or equilibrium, the magnetization vanishes; consequently, the long time limit of the time correlation function $\langle M(t)\tilde{M}\rangle_{eq}$ is zero. The final result is

$$\chi(\omega) = \frac{\beta}{V}\langle M\tilde{M}\rangle_{eq} - \frac{\beta}{V}i\omega\int_0^\infty dt\, \exp(-i\omega t)\langle M(t)\tilde{M}\rangle_{eq}. \quad (7.69)$$

The zero frequency part of this equation is the familiar relation between equilibrium magnetic susceptibility and fluctuations in the magnetic moment. If the magnetic dipole correlation function decays exponentially,

$$\langle M(t)\tilde{M}\rangle_{eq} = \langle M\tilde{M}\rangle_{eq}\exp(-t/\tau), \quad (7.70)$$

then the frequency-dependent susceptibility has the familiar form,

$$\chi(\omega) = \frac{\beta}{V}\langle M\tilde{M}\rangle_{eq}\frac{1}{1+i\omega\tau}. \quad (7.71)$$

Any more-complicated frequency dependence is a clear indication that the magnetic dipole time correlation function has a more-complicated time dependence. This is an example of how spectral information can provide dynamical information.

The last application to be discussed now is the theory of energy absorption in an electric field. If the perturbation is the interaction of an applied field $E(t) = E_0\cos(\omega t)$ with an electric dipole moment M (the component in the direction of the field), and the response is the electric current $J = dM/dt$, we obtain Ohm's law,

$$\langle J; t\rangle = \int_0^t ds\, \phi(s)E(t-s) + \cdots. \quad (7.72)$$

The response function is

$$\phi(t) = \beta\langle J(t)\tilde{J}\rangle. \quad (7.73)$$

On expanding the cosine, we find

$$\langle J; t\rangle = E_0\cos\omega t\int_0^t ds\, \cos\omega s\,\phi(s) + E_0\sin\omega t\int_0^t ds\, \sin\omega s\,\phi(s). \quad (7.74)$$

We expect that the response function decays to zero at long times and is integrable. Then, at long times the upper limits of the two integrals can be replaced by infinity,

$$\langle J;t\rangle \to E_0 \cos\omega t \int_0^\infty ds\, \cos\omega s\, \phi(s) + E_0 \sin\omega t \int_0^\infty ds\, \sin\omega s\, \phi(s). \tag{7.75}$$

The rate of energy dissipation (Joule heating) at time t is $\langle J;t\rangle E(t)$. We average this over a long time; the sine term drops out and the cosine term gives a factor of $1/2$,

$$(\dot{E})_{\text{diss}} = \frac{1}{2} E_0^2 \int_0^\infty ds\, \cos\omega s\, \phi(s). \tag{7.76}$$

This is the energy absorbed by the system per unit time. Recall that an earlier section presented another quite different calculation of the same quantity. The two results do not appear to be the same, but they are, once one takes account of some important identities relating various time correlation functions in quantum mechanics.

Some Identities

The derivation of these identities begins with the symmetrized correlation function

$$C(t) = \frac{1}{2}\langle AA(t) + A(t)A\rangle. \tag{7.77}$$

This particular correlation function is symmetric in t, $C(-t) = C(t)$. To see this, we note that by shifting the time origin, $\langle AA(-t)\rangle = \langle A(t)A\rangle$ and $\langle A(-t)A\rangle = \langle AA(t)\rangle$. Then the definition of $C(t)$ leads directly to the time symmetry. This correlation function has the spectral density (defined by the Fourier transform)

$$C_\omega = \int_{-\infty}^\infty dt\, e^{-i\omega t} C(t). \tag{7.78}$$

When written out in detail in the energy representation with the abbreviation $\omega_{mn} = (E_m - E_n)/\hbar$, $C(t)$ is

$$C(t) = \frac{1}{2}\sum_{m,n} \rho_m \langle m|A|n\rangle\langle n|A|m\rangle (e^{i\omega_{mn}t} + e^{i\omega_{nm}t}). \tag{7.79}$$

The Fourier integration produces two delta functions,

$$C_\omega = \pi \sum_{m,n} \rho_m \langle m|A|n\rangle\langle n|A|m\rangle (\delta(\omega - \omega_{mn}) + \delta(\omega + \omega_{mn})). \quad (7.80)$$

We can eliminate the second delta function by switching m and n in the second term,

$$C_\omega = \pi \sum_{m,n} (\rho_m + \rho_n)\langle m|A|n\rangle\langle n|A|m\rangle \delta(\omega - \omega_{mn}). \quad (7.81)$$

Because ρ is a Boltzmann distribution, we can write

$$\frac{\rho_m}{\rho_n} = e^{-\beta\hbar\omega_{mn}}. \quad (7.82)$$

The delta function allows us to replace ω_{mn} by ω. We substitute in the preceding equation,

$$C_\omega = \pi(1 + e^{-\beta\hbar\omega}) \sum_{m,n} \rho_m \langle m|A|n\rangle\langle n|A|m\rangle \delta(\omega - \omega_{mn}). \quad (7.83)$$

But this is proportional to the Fourier transform of $\langle AA(t)\rangle$,

$$C_\omega = \frac{1}{2}(1 + e^{-\beta\hbar\omega}) \int_{-\infty}^{\infty} dt\, e^{-i\omega t} \langle AA(t)\rangle. \quad (7.84)$$

The same reasoning leads to two more identities,

$$C_\omega = \frac{1}{2}(e^{\beta\hbar\omega} + 1) \int_{-\infty}^{\infty} dt\, e^{-i\omega t} \langle A(t)A\rangle \quad (7.85)$$

and

$$\int_{-\infty}^{\infty} dt\, e^{-i\omega t} \langle A(t)A\rangle = e^{-\beta\hbar\omega} \int_{-\infty}^{\infty} dt\, e^{-i\omega t} \langle AA(t)\rangle. \quad (7.86)$$

Because $C(t)$ is even in t, its spectral density is real, even in ω, and is a cosine transform,

$$C_\omega = \int_{-\infty}^{\infty} dt\, e^{-i\omega t} C(t) = 2\int_0^{\infty} dt\, \cos\omega t\, C(t). \quad (7.87)$$

Next, we consider the relation of the symmetrized correlation function to the one that is generated by Kubo theory,

$$C_K(t) = \langle A(t)\tilde{A}\rangle = \frac{1}{\beta}\int_0^\beta d\lambda \langle A(t)A(i\hbar\lambda)\rangle. \quad (7.88)$$

The Fourier transform of this function is

$$\int_{-\infty}^{\infty} dt\, e^{-i\omega t} C_K(t) = \frac{2\pi}{\beta} \int_0^{\beta} d\lambda \sum_{m,n} p_m \langle m|A|n\rangle \langle n|A|m\rangle \delta(\omega - \omega_{mn}) e^{\lambda\hbar\omega_{mn}}. \tag{7.89}$$

The delta function allows us to change the exponent from ω_{mn} to ω. The λ integral is easy, and the remaining sum can be converted back to a Fourier integral,

$$= 2\pi \frac{e^{\beta\hbar\omega} - 1}{\beta\hbar\omega} \int_{-\infty}^{\infty} dt\, e^{-i\omega t} \langle A(t)A\rangle. \tag{7.90}$$

On using one of the earlier identities, eq. (7.85), we can finally write

$$\int_{-\infty}^{\infty} dt\, e^{-i\omega t} C_K(t) = \frac{2}{\beta\hbar\omega} \frac{1 - e^{-\beta\hbar\omega}}{1 + e^{-\beta\hbar\omega}} C_\omega. \tag{7.91}$$

The right-hand side is real and even in ω, so the left-hand side is the cosine transform,

$$2\int_0^{\infty} dt\, \cos\omega t\, C_K(t) = \frac{2}{\beta\hbar\omega} \frac{1 - e^{-\beta\hbar\omega}}{1 + e^{-\beta\hbar\omega}} C_\omega. \tag{7.92}$$

By applying the identity eq. (7.84) and canceling, this is converted to

$$\int_0^{\infty} dt\, \cos\omega t\, C_K(t) = \frac{1 - e^{-\beta\hbar\omega}}{2\beta\hbar\omega} \int_{-\infty}^{\infty} dt\, e^{-i\omega t} \langle AA(t)\rangle. \tag{7.93}$$

Let us return to the calculation of energy absorption. For this application, we replace A by the current J,

$$\int_0^{\infty} dt\, \cos\omega t\, \langle J(t)\tilde{J}\rangle = \frac{1 - e^{-\beta\hbar\omega}}{2\beta\hbar\omega} \int_{-\infty}^{\infty} dt\, e^{-i\omega t} \langle JJ(t)\rangle. \tag{7.94}$$

Then, according to the earlier calculation, the energy absorption is

$$(\dot{E})_{\text{abs}} = \frac{1}{2} E_0^2 \int_0^{\infty} dt\, \cos\omega t\, \beta \langle J(t)\tilde{J}\rangle$$

$$= \frac{1 - e^{-\beta\hbar\omega}}{4\hbar\omega} E_0^2 \int_{-\infty}^{\infty} dt\, e^{-i\omega t} \langle JJ(t)\rangle. \tag{7.95}$$

Finally, we note that in Fourier integrals, a time derivative is equivalent to a factor $i\omega$, and $J = dM/dt$, so that

$$\int_{-\infty}^{\infty} dt\, e^{-i\omega t} \langle JJ(t) \rangle \to \omega^2 \int_{-\infty}^{\infty} dt\, e^{-i\omega t} \langle MM(t) \rangle. \tag{7.96}$$

Then the resulting energy absorption is identical with the formula derived in an earlier section by means of the Golden Rule,

$$(\dot{E})_{\text{abs}} = \frac{\omega}{4\hbar}(1 - e^{-\beta\hbar\omega})\int_{-\infty}^{\infty} dt\, e^{-i\omega t} \langle MM(t) \rangle E_0^2. \tag{7.97}$$

8

Projection Operators

8.1 Projection Operators and Hilbert Space

Theoretical treatments of nonequilibrium systems are often based on master equations or Langevin or Fokker-Planck equations. When we use such models, we typically do not hope for a complete, detailed, and exact treatment of a problem. Rather, we tend to think in terms of approximations in which irrelevant details are omitted and only those aspects of the problem that appear to be physically important are included. This is a standard approach to the theoretical analysis of complex situations; when it is used with common sense, it is a very productive one.

But in constructing approximate models, we should always keep in mind certain questions. Do we know that the approximate model has a more exact statistical mechanical basis? For example, can a hypothesized nonlinear Langevin equation actually be obtained from a well-defined Hamiltonian? Do we know how to improve on a chosen approximate model? That is, do we know what was left out and how to put it in if we wish? Projection operator methods provide some answers.

We have already seen how projection operators can be used to derive quantum mechanical master equations (section 6.5). This section provides an introduction to projection operators in a form that is motivated by their use in deriving Langevin equations. The following sections provide a formal derivation of a non-Markovian linear Langevin equation for a set of dynamical variables, a demonstration that it reduces to a Markovian Langevin equation if the variables are "slow," and some understanding of the limits of validity of linear Langevin equations. The results are essentially the equations that were introduced in section 1.4. Later chapters will deal with nonlinear Langevin equations and their corresponding Fokker-Planck equations.

Matrix Form of the Liouville Equation

The Liouville equation for the dynamical variable $A(t)$,

$$\frac{\partial}{\partial t} A(t) = LA(t), \qquad (8.1)$$

is a linear differential equation in A. The linearity suggests constructing a matrix representation of the dynamics.

Any dynamical quantity $A(X)$ can be expanded in an infinite set of functions $\varphi_j(X)$ in the Hilbert space of all functions of X. This is like what one customarily does in quantum mechanics. Each function φ_j is like a vector in this Hilbert space. Then A is itself a vector in Hilbert space.

In order to make expansions, we need a rule for forming the inner product (or dot product) of two vectors A and B. The inner product is denoted in a nonspecific way by (A, B). The actual rule to be used may depend on circumstances in a way that will be discussed later. However, it may help the reader to keep in mind one particular choice, because it turns out to be a common one,

$$(A, B) = \int dX\, f_{eq}(X) A(X) B^*(X) = \langle AB^* \rangle_{eq}, \qquad (8.2)$$

where * denotes the complex conjugate. Once an inner product has been selected, then the set $\varphi_j(X)$ can be constructed so that its individual vectors are orthogonal and normalized,

$$(\varphi_j, \varphi_k) = \delta_{jk}. \qquad (8.3)$$

Now we can expand any time-dependent dynamical variable $A(X,t)$ in this orthonormal set,

$$A(X,t) = \sum_m a_m(t)\varphi_m(X), \qquad (8.4)$$

where the coefficients are

$$a_m(t) = (A(t), \varphi_m). \qquad (8.5)$$

PROJECTION OPERATORS

When the expansion is put into the Liouville equation, the resulting equation takes on the simple matrix-vector form,

$$\frac{\partial}{\partial t} a_m(t) = \sum_n L_{mn} a_n(t), \tag{8.6}$$

where the Liouville operator is replaced by its matrix representation,

$$L_{mn} = (\varphi_m, L\varphi_n). \tag{8.7}$$

If the explicit inner product in eq. (8.2) is used, then the matrix L_{mn} is anti-Hermitian, and the exponential operator $\exp(tL)$ is unitary. The effect of a unitary operator on any vector in Hilbert space is to rotate it without changing its magnitude.

A similar expansion can be made to convert the Liouville equaton for a phase space distribution,

$$\frac{\partial}{\partial t} f(X,t) = -Lf(X,t), \tag{8.8}$$

into a vector-matrix equation. The expansion is

$$f(X,t) = f_{eq}(X) \sum_j b_j(t) \varphi_j(X). \tag{8.9}$$

The coefficient $b_j(t)$ is the time-dependent average of the function $\varphi_j(X)$,

$$b_j(t) = \int dX \varphi_j(X) f(X,t), \tag{8.10}$$

and obeys the matrix equation

$$\frac{\partial}{\partial t} b_j(t) = -\sum_k L_{jk} b_k(t) \tag{8.11}$$

when the special inner product of eq. (8.2) is used. The matrix L_{jk} is the same as in eq. (8.7).

All of the above is exactly what one does in quantum mechanics, except that the Liouville operator is a first-order differential operator and the Hamiltonian operator is second order.

There is usually not much practical advantage in going from a representation involving partial differential equations to another representation involving infinite matrices. However, the expansion suggests a way of finding Langevin equations for macroscopic dynamical variables. The process works because we are usually interested in only a small subset of the vectors that define the whole Hilbert space.

The set of dynamical variables $\{A_j(X)\}$ in the Langevin equation are vectors in Hilbert space. They are usually not orthonormal, but this can be fixed by the Gram-Schmidt process. First we normalize A_1; this gives one unit vector. Then we select A_2, subtract enough of A_1 to make a vector that is orthogonal to A_1, and normalize it. This gives a second unit vector, orthogonal to the first. We continue this process of successive subtraction, orthogonalization, and normalization, until all members of the set are accounted for. The result is an orthonormal set corresponding to $\{A_j\}$, and along with it, a subspace of the complete Hilbert space. Any linear combination of the As lies in this subspace.

Why should we go through all this? The reason is that macroscopic equations of motion are approximately self-determined. In hydrodynamics, for example, the density, temperature, and fluid velocity at time t are determined by the same quantities at an earlier time. Details of individual molecular motion are not relevant. Following this example, we hope more generally that the dynamical behavior of any chosen set $\{A\}$ will be concentrated in the subspace spanned by $\{A\}$. We hope that the initial values of variables that are orthogonal to the chosen set are unimportant. In this hope, we call members of the set $\{A\}$ "relevant" dynamical variables and variables orthogonal to the set "irrelevant." Statistical mechanics does not tell us what the relevant variables are. This is our choice. If we choose well, the results may be useful; if we choose badly, the results (while still formally correct) will probably be useless.

Of course, this hoped-for deterministic behavior does not really happen. While by definition A starts out in the relevant subspace, in time the Hilbert space rotation generated by $\exp(tL)$ will take $A(t)$ out of this subspace, so that it picks up contributions from the initial values of the irrelevant variables. This gives rise to the noise in the Langevin equation.

Partitioning

One way to focus on the dynamics of a particular subset of all dynamical variables, used already in the derivation of the generalized master equation, is to partition the Liouville matrix. This can be done formally with projection operators. First, some mathematically simple examples are used to show how the general problem is going to be treated.

By far the simplest nontrivial problem is two-dimensional. Then any dynamical variable is represented by the two component vector (a_1, a_2). The Liouville equation becomes a pair of linear equations for the two coefficients,

$$\frac{\partial}{\partial t}\begin{pmatrix} a_1 \\ a_2 \end{pmatrix} = \begin{pmatrix} L_{11} & L_{12} \\ L_{21} & L_{22} \end{pmatrix} \cdot \begin{pmatrix} a_1 \\ a_2 \end{pmatrix}. \tag{8.12}$$

Suppose that we are interested in only the coefficient a_1, which will be referred to as "relevant," and we do not care what the other "irrelevant" coefficient a_2 is doing. Then we solve the second of these equations for a_2,

$$a_2(t) = \exp(L_{22}t)a_2(0) + \int_0^t ds\, \exp(L_{22}(t-s))L_{21}a_1(s), \tag{8.13}$$

and we put this back into the equation for a_1,

$$\frac{\partial}{\partial t}a_1(t) = L_{11}a_1(t) + L_{12}\int_0^t ds\, \exp(L_{22}(t-s))L_{21}\, a_1(s)$$
$$+ L_{12}\exp(L_{22}t)a_2(0). \tag{8.14}$$

The result of this rearrangement is an equation of motion for $a_1(t)$ in which the other coefficient a_2 appears only as an initial value $a_2(0)$. If this initial value vanishes, or if we are willing to ignore its effects, regarding them as irrelevant "noise," then the relevant coefficient $a_1(t)$ satisfies its own equation. (As an example, it may happen that we are only interested in the average of a_1 over some initial distribution of a_2 and that the average of the initial a_2 vanishes.) Note that this equation is now non-Markovian.

Exactly the same procedure can be used for a Hilbert space of N dimensions. We may treat the first n coefficients as a vector a_1 in the relevant subspace of n dimensions, and the remaining $N-n$ coefficients as the vector a_2 in the orthogonal or irrelevant subspace. Then L_{11} is a matrix of order $n \times n$, L_{12} is of order $n \times (N-n)$, L_{21} is of order $(N-n) \times n$, and L_{22} is of order $(N-n) \times (N-n)$. The preceding scalar equation has been written so that it is still correct with this vector reinterpretation of the symbols. The rate of change of a_1 has two parts; one is the effect of the history of a_1 back to the initial time, and the other is the effect of the initial conditions of the irrelevant coefficients a_2, or "noise." Clearly, elimination of variables transforms a Markovian (i.e., no memory) equation into a non-Markovian equation. Also, the effects of the eliminated irrelevant variables are very much like the noise in Langevin equations.

Projecting

The procedure just described involved *partitioning* of a vector and matrix. This can be done in mathematical notation by introducing a matrix projection operator,

$$\mathbf{P} = \begin{pmatrix} 1 & 0 \\ 0 & 0 \end{pmatrix} = \begin{pmatrix} n \times n & n \times (N-n) \\ (N-n) \times n & (N-n) \times (N-n) \end{pmatrix}, \quad (8.15)$$

where the second matrix is used solely to indicate the dimensionality of the partitions of the first matrix. Note that this matrix is idempotent; that is, it obeys the condition $\mathbf{PP} = \mathbf{P}$, which is a general requirement for a projection matrix. Then a more abstract vector-matrix equation for **a** is

$$\frac{\partial}{\partial t}\mathbf{a} = \mathbf{L} \cdot \mathbf{a}, \quad (8.16)$$

and the partitioned vector and matrices are

$$\mathbf{a}_1 = \mathbf{P} \cdot \mathbf{a}, \quad \mathbf{a}_2 = (1-\mathbf{P}) \cdot \mathbf{a}, \quad \mathbf{L}_{11} = \mathbf{P} \cdot \mathbf{a} \cdot \mathbf{P}, \quad \mathbf{L}_{12} = \mathbf{P} \cdot \mathbf{L} \cdot (1-\mathbf{P})$$
$$\mathbf{L}_{21} = (1-\mathbf{P}) \cdot \mathbf{L} \cdot \mathbf{P}, \quad \mathbf{L}_{22} = (1-\mathbf{P}) \cdot \mathbf{L} \cdot (1-\mathbf{P}).$$
$$(8.17)$$

In this form, all vectors still have dimension N, although $\mathbf{a}_1$ has $(N-n)$ zero elements and $\mathbf{a}_2$ has n zero elements. Similar statements apply to the matrices, all of which are $N \times N$. There is no problem in reinterpreting eqs. (8.13) and (8.14) in the full space.

The Subspace of the Relevant Variables

The systems we are concerned with generally involve an infinite-dimensional Hilbert space. The subspace of interest is spanned by relevant variables. The orthogonal part of the whole space is spanned by irrelevant variables. Projection onto the relevant subspace is a partitioning of the sort just discussed. While the projection can be specified by constructing all the relevant unit vectors, it is easier to specify it by the abstract operator **P**. Given any set $\{A\}$, the projection operator acting on any variable B is given explicitly by

$$\mathbf{P}B = (B, A) \cdot (A, A)^{-1} \cdot A, \quad (8.18)$$

where the inner product (B, A) has the dimensionality of the vector A, and the inner product (A, A) is an $n \times n$ matrix. The ordering of symbols in this formula is designed so that it is easy to put in subscripts if desired:

$$\mathbf{P}B = \sum_j \sum_k (B, A_j)\left((A, A)^{-1}\right)_{jk} A_k. \quad (8.19)$$

If the variables $\{A\}$ have already been orthonormalized, then (A, A) is the identity matrix. Note that the inner products are numbers and not dynamical variables. It is easy to verify that **P** is a projection, that is, **PP** = **P**. If B lies in the subspace spanned by $\{A\}$, then **P**B also lies in that subspace and is the same as B.

It is especially important, and fortunate, that we need to specify only the relevant variables in order to make a projection onto the relevant subspace. We don't need to take on the task of enumerating all the irrelevant variables. The orthogonal part of Hilbert space, with its infinity of coordinates, is selected by the operator $(1 - \mathbf{P})$.

Another somewhat different example of partitioning or projecting was used in the derivation of the generalized master equation. There, the density matrix was partitioned into a relevant diagonal part and an irrelevant off-diagonal part.

8.2 Derivation of Generalized Langevin Equations

This section presents a derivation of generalized Langevin equations (H. Mori, 1965) of the form that was briefly mentioned in section 1.4,

$$\frac{da(t)}{dt} = i\Omega\, a(t) - \int_0^t ds K(s) a(t-s) + F(t). \qquad (8.20)$$

The starting point is the Liouville equation, and the resulting eq. (8.20) can be regarded as a mathematical rearrangement of the Liouville equation. The derivation to be given here is based on abstract operator manipulations (J. T. Hynes and J. M. Deutch, 1975) that were designed to get to the desired result as quickly as possible. First we separate the Liouville operator into two parts,

$$L = \mathbf{P}L + (\mathbf{1} - \mathbf{P})L. \qquad (8.21)$$

Next we use an operator identity,

$$e^{tL} = e^{t(1-\mathbf{P})L} + \int_0^t ds\, e^{(t-s)L} \mathbf{P} L e^{s(1-\mathbf{P})L}. \qquad (8.22)$$

This can be verified either directly by differentiation or indirectly by taking Laplace transforms and using the convolution theorem. Next, we operate with both sides of this equation on the quantity $(1 - \mathbf{P})\,L\,A$. On the left-hand side, we get

$$e^{tL}(\mathbf{1}-\mathbf{P})LA = e^{tL}LA - e^{tL}\mathbf{P}LA$$

$$= \frac{\partial}{\partial t} e^{tL} A - e^{tL}(LA, A)\cdot(A, A)^{-1}\cdot A$$

$$= \frac{\partial}{\partial t} A(t) - (LA, A)\cdot(A, A)^{-1}\cdot A(t). \qquad (8.23)$$

In the last line, we use $A(t) = \exp(tL) A$, and we recognize that the inner products are numbers that commute with the operator $\exp(tL)$.

On the right-hand side, we define

$$F(t) = e^{t(1-\mathbf{P})L}(\mathbf{1} - \mathbf{P})LA; \qquad (8.24)$$

this will turn out to be the "noise" term in the Langevin equation. Then the right-hand side becomes

$$F(t) + \int_0^t ds\, e^{(t-s)L}(LF(s), A) \cdot (A, A)^{-1} \cdot A$$
$$= F(t) + \int_0^t ds\, (LF(s), A) \cdot (A, A)^{-1} \cdot A(t-s). \qquad (8.25)$$

Finally, we define the matrices Ω and $\mathbf{K}$ by

$$i\Omega = (LA, A) \cdot (A, A)^{-1} \qquad (8.26)$$

$$\mathbf{K}(t) = -(LF(t), A) \cdot (A, A)^{-1}. \qquad (8.27)$$

If the inner product is such that L is anti-Hermitian, then this is also

$$\mathbf{K}(t) = (F(t), LA) \cdot (A, A)^{-1} = (e^{t(1-\mathbf{P})L}(\mathbf{1} - \mathbf{P})LA, LA) \cdot (A, A)^{-1}.$$
$$\qquad (8.28)$$

The result of this formal algebraic manipulation is

$$\frac{\partial}{\partial t} A(t) = i\Omega \cdot A(t) - \int_0^t ds\, \mathbf{K}(s) \cdot A(t-s) + F(t). \qquad (8.29)$$

This is, by construction, a rearrangement of the original Liouville equation; it is a mathematical identity without immediate physical meaning. However, it surely looks like the Langevin equations discussed earlier. The first two terms on the right-hand side describe the systematic part of the motion of the chosen relevant variables, as determined by their initial values, and the third term gives the effects of the initial values of the irrelevant variables.

There are subtle differences between this equation and the exact Langevin equation derived earlier for a system interacting with a heat bath of harmonic oscillators. The position and velocity of the system may be taken as the relevant variables, and all the heat bath coordinates and momenta as irrelevant. The present derivation leads to a generalized Langevin equation that is inherently *linear* in the system variables. The earlier derivation led in general to a *nonlinear* Langevin equation. Only if the system Hamiltonian is quadratic in position do the two equations agree. The reason for the disagreement in general, and the agreement in the case of a harmonic system, lies in the difference between phase space and the Hilbert space of dynamical variables. In phase space, we do not regard x and x^2 as different dynamical variables: If we know x, then we know x^2. But in Hilbert space, x and x^2 are different vectors, not necessarily parallel. They contribute differently to the evolution of any $A(X; t)$. If the system and heat bath are both harmonic, then the Liouville operator converts any linear combination of coordinates and momenta into another linear combination; the subspace of linear combinations of the basic variables is dynamically closed. There is no reason to distinguish between phase space and the linear subspace. If, however, the system is not harmonic, the Liouville operator takes one out of this linear subspace. Then the two Langevin equations become different.

Up to this point, the derivation did not discriminate between classical mechanics and quantum mechanics. The Liouville operator could be classical or quantum. The derivation involved only some symbolic manipulation of operators and variables. Also, the exact nature of the inner product (A, B) was never mentioned. What finally determines what inner product to use is the requirement that the average of the noise must be as small as possible. The following section deals with the treatment of noise.

8.3 Noise in Generalized Langevin Equations

If the equation just derived is to be regarded as a conventional Langevin equation and not just as a mathematical identity, then the quantity $F(t)$ ought to have the properties that are expected of Langevin noise. In particular, its average over some initial nonequilibrium distribution should vanish. So we digress briefly on the construction of initial nonequilibrium distribution functions.

Initial Nonequilibrium States

We know how to treat thermal equilibrium in statistical mechanics; we generally use the Gibbs distribution function or density matrix. This is easy because there is only one thermal equilibrium state. (Microcanonical, canonical, and grand canonical ensemble distribution functions all give rise to the same macroscopic thermodynamics; they differ only in the kinds of fluctuations that are allowed.) But there are many possible nonequilibrium states. How can we find the appropriate distribution function for any given one of these? This is a central problem of nonequilibrium statistical mechanics, and it is a hard question to answer in a general way. Two simple recipes are commonly used, one of which is operationally sound but limited in applicability, and the

152 NONEQUILIBRIUM STATISTICAL MECHANICS

other more generally applicable but not so well justified. Fortunately, the two recipes lead to similar conclusions.

One approach is to start with an equilibrium state, with a Hamiltonian H and canonical distribution function,

$$f_{eq} = \frac{1}{Q}\exp(-\beta H), \qquad (8.30)$$

and switch on an constant external field E with the same perturbing Hamiltonian $-ME$ that was used in the last chapter. (Again we assume that the equilibrium average of M vanishes.) We wait long enough for the system to come to a new equilibrium state in the presence of this field. Then at $t = 0$, we switch off the field. Now the system has once more the original unperturbed Hamiltonian H but is in a state described by the nonequilibrium distribution,

$$f(X;0) = \frac{1}{Q(E)}\exp(-\beta H + \beta ME). \qquad (8.31)$$

The average of M in this initial state is

$$\langle M;0\rangle = \frac{\partial \log Q(E)}{\partial \beta E} = \chi_{MM} E + O(E^2). \qquad (8.32)$$

This can be solved to replace the field strength E by the initial average,

$$E = \frac{1}{\chi_{MM}}\langle M;0\rangle + O(\langle M;0\rangle^2). \qquad (8.33)$$

In all that follows, we are concerned only with small deviations from equilibrium, so the quadratic term will be neglected. Then the initial nonequilibrium distribution, to first order in deviation from equilibrium, is

$$f(X;0) = f_{eq}(X)(1 + \beta M(X)E + \cdots)$$
$$= f_{eq}(X)\left(1 + \beta\frac{\langle M;0\rangle}{\chi_{MM}}M(X) + \cdots\right). \qquad (8.34)$$

Having an initial state, we can now determine the time dependence of the average of any variable A. (We assume that its equilibrium average vanishes.) The average, to first order in deviation from equilibrium, is

$$\langle A;t\rangle = \int dX A(t;X)f(X;0) = \beta\frac{\langle M;0\rangle}{\chi_{MM}}\langle A(t)M(0)\rangle_{eq} + \cdots. \qquad (8.35)$$

The relaxation of $\langle A; t\rangle$ back to its equilibrium value follows the time correlation function of A and M. In particular, we find for $A = M$,

$$\frac{\langle M; t\rangle}{\langle M; 0\rangle} = \frac{\langle M(t)M\rangle_{eq}}{\langle MM\rangle_{eq}} + \cdots. \tag{8.36}$$

This construction of an initial nonequilibrium state corresponds to a well-defined experimental process. But it is limited to situations where we have at our disposal an external field and as was noted earlier, there are very few fields that are experimentally available. We can always assume the existence of a hypothetical field that couples to some arbitrarily chosen variable M and proceed with that. Because this procedure cannot be carried out in a real experiment, we have no way of knowing whether it is reliable.

Another approach that is commonly used relies on a maximum entropy argument. This is one of the standard textbook ways of justifying the Gibbs ensemble. The procedure is to start with the Boltzmann entropy,

$$S = k_B \int dX f(X) \log f(X), \tag{8.37}$$

impose two constraints,

$$\int dX f(X) = 1, \qquad \int dX H(X) f(X) = U, \tag{8.38}$$

where U is some assigned energy, and then maximize the entropy subject to the constraints. The variational calculation introduces two Lagrange multipliers α and β, and the maximum entropy solution is

$$f(X) = \exp(-\alpha - \beta H(X)). \tag{8.39}$$

The normalization constraint leads to

$$\exp(\alpha) = Q(\beta) = \int dX \exp(-\beta H(X)), \tag{8.40}$$

and the energy constraint leads to

$$U = -\frac{\partial \log Q}{\partial \beta}, \tag{8.41}$$

which should be inverted to give β as a function of the assigned U. All of this is familiar equilibrium statistical mechanics. Perhaps the best argument in favor of the maximum entropy method is that it leads to correct results.

Now we add a further constraint, that the average of some variable M has the assigned value m. Then the maximum entropy calculation requires another Lagrange multiplier, which we choose to denote by βE, and

$$(f(X))_{\max} = \exp(-\alpha - \beta H(X) + \beta E M(X)). \tag{8.42}$$

Just as before, the constraints lead to

$$\exp(\alpha) = Q(\beta, E) = \int dX \exp(-\beta H(X) + \beta E M(X)), \tag{8.43}$$

$$U = -\frac{\partial \log Q}{\partial \beta}, \qquad m = \frac{\partial \log Q}{\partial \beta E} \tag{8.44}$$

Then β and E are functions of the assigned U and m.

If the chosen variable M is a constant of the motion, then this procedure leads to a good equilibrium distribution. But generally it is not, and then the resulting distribution cannot be stationary in time. Whether this maximum entropy distribution can actually be a valid initial distribution for a real nonequilibrium experiment is not certain. The maximum entropy argument suggests, however, that it is as good a guess as we are likely to make. Further, when the maximum entropy argument is applied to the case where we can make an initial nonequilibrium state by applying an external field, the result is the same.

This procedure can be used to construct a quite general initial nonequilibrium distribution. We start with a chosen set of dynamical variables $\{A_j, j = 1, 2, \ldots\}$. For generality, we allow complex quantities and add complex conjugates in the appropriate places. Their equilibrium averages all vanish. If we assign average values $\{a_j\}$ to these variables, the maximum entropy distribution is

$$f = f_{eq}\left\{1 + \sum_j \gamma_j A_j^* + O(\gamma^2)\right\}, \tag{8.45}$$

and the Lagrange multipliers $\{\gamma_j\}$ are determined by the assigned averages,

$$a_j = \langle A_j; 0 \rangle = \sum_k \langle A_j A_k^* \rangle_{eq} \gamma_k + O(\gamma^2). \tag{8.46}$$

To save space we abbreviate

$$\mathbf{M} = \langle AA^* \rangle_{eq} \qquad \text{or} \qquad M_{jk} = \langle A_j A_k^* \rangle_{eq}. \tag{8.47}$$

This can be solved for the γs by inverting the matrix $\mathbf{M}$. In vector-matrix form, the solution is

$$\gamma = \mathbf{M}^{-1} \cdot a + O(a^2). \tag{8.48}$$

Time-dependent averages of these variables can be found by multiplying the initial distribution by $A(t)$ and integrating,

$$\langle A; t \rangle = \langle A(t) A^* \rangle \cdot \mathbf{M}^{-1} \cdot a + \cdots \tag{8.49}$$

to lowest order in deviation from equilibrium. Again we see how the nonequilibrium averages follow the time dependence of the equilibrium time correlation functions (Onsager's regression).

All of this discussion has been based on classical statistical mechanics. The quantum version differs only in the way one expands the exponential of two noncommuting operators. As in the previous treatment of linear response to an external field, the only change that must be made is to replace the variables $\{A\}$ in the expansion by their Kubo transforms,

$$\rho = \left\{ 1 + \sum_j \gamma_j \tilde{A}_j^* + \cdots \right\} \rho_{eq}. \tag{8.50}$$

Averaged Noise

Now we can return to the properties of noise in the formally exact Langevin equation. The hope is that the average of the noise $\langle F(t) \rangle$ vanishes or at least is negligibly small. The average is taken over a statistical ensemble of initial conditions that is not at equilibrium but close to equilibrium. Without any further discussion, we use the maximum entropy distribution,

$$f(X; 0) = f_{eq}(X) \left\{ 1 + \sum_j \gamma_j A_j^*(X) + O(\gamma^2) \right\} \tag{8.51}$$

from eq. (8.45). The Lagrange multipliers are connected to the initial averages (in vector-matrix form) by

$$\langle A; 0 \rangle = \langle AA^* \rangle_{eq} \cdot \gamma + O(\gamma^2)$$
$$\gamma = \langle AA^* \rangle_{eq}^{-1} \cdot \langle A; 0 \rangle + O(\langle A; 0 \rangle^2). \tag{8.52}$$

Having chosen an initial nonequilibrium ensemble, now we can average the Langevin equation over the initial state,

$$\frac{\partial}{\partial t} \langle A; t \rangle = i\Omega \cdot \langle A; t \rangle - \int_0^t ds \mathbf{K}(s) \cdot \langle A(t-s) \rangle + \langle F(t) \rangle. \tag{8.53}$$

If the future behavior of $\langle A; t \rangle$ is determined by its present and earlier values, and not by irrelevant variables, then the average of $F(t)$ over the initial ensemble must be negligible in comparison with the other terms in the equation.

The average of $F(t)$ over initial conditions is, to first order in γ,

$$\langle F(t)\rangle = \int dX F(t) f_{eq}(X)\{1+\gamma \cdot A^*(X)+O(\gamma^2)\}. \tag{8.54}$$

The first term, which is independent of γ, is the equilibrium average of $F(t)$ and must vanish. Since an initial equilibrium state stays in equilibrium for all time, we can replace $\langle A; 0\rangle$ by $\langle A\rangle_{eq} = 0$ everywhere in the averaged equation, and so the equilibrium average of the remaining $F(t)$ must also vanish.

The second term is of order γ and is therefore first order in the initial deviation from equilibrium. The other terms of the averaged Langevin equation, that contain $\langle A; t\rangle$ are also first order in deviation from equilibrium. To have a sensible Langevin equation, at least to this order, we must somehow arrange that the $O(\gamma)$ term in the average of $F(t)$ will vanish, or

$$\int dX f_{eq}(X) F(t) A^*(X) = 0. \tag{8.55}$$

Up to this point, even though the quantities $i\,\Omega$, $\mathbf{K}(t)$, and $F(t)$ all involve the chosen inner product, we did not need to know what it was. *Now we can force this average to vanish by making the right choice of inner product.* It is evident that the inner product that was suggested earlier,

$$(A, B) = \int dX f_{eq}(X) A(X) B^*(X), \tag{8.56}$$

will work. To verify this, note that if we make this choice, the integral in eq. (8.55) is just the inner product of $F(t)$ and A. But by construction, $F(t)$ lies entirely in the subspace that is orthogonal to $\{A\}$; if $F(t)$ is expanded in powers of t, every term in the expansion contains the factor $(1 - \mathbf{P})$. With this particular choice of inner product, the average of $F(t)$ vanishes automatically to first order in γ.

The result is that $\langle F(t)\rangle$ is of order γ^2. The average of the "noise" is second order in deviations from equilibrium, while the other terms in the averaged Langevin equation are all first order in deviations from equilibrium. Thus the averaged noise is negligible for an initial state that is close enough to equilibrium. In this sense, the exact eq. (8.39) in the preceding chapter can be used as an approximate Langevin equation. This use, however, is restricted to linear or near-equilibrium transport processes. In treating nonlinear processes, $\langle F(t)\rangle$ cannot be neglected; the separation into "systematic" and "noisy" terms is not under control.

Quantum Mechanics

The preceding treatment of noise was based on classical statistical mechanics. As in the discussion of linear response to an external force, the only significant change in quantum mechanics is in the use of the Kubo transform. Once we have chosen an inner product, replaced the classical Liouville equation by its quantum form, and replaced the phase space distribution function by the density matrix, the derivation is exactly the same as with classical mechanics. To average the noise, we need an initial nonequilibrium density matrix. Its first-order term contains the Kubo transform of A. When we average the noise over this initial state, we encounter the quantity $\langle F(t)\tilde{A}^*\rangle_{eq}$ in first order. To make this vanish, we must define the quantum mechanical inner product by

$$(A, B)_{QM} = \text{Trace} A \cdot \tilde{B}^* \cdot \rho_{eq}. \tag{8.57}$$

Then the average noise is second order in deviation from equilibrium. The QM inner product must be used in calculating $i\Omega$ and $\mathbf{K}$.

8.4 Generalized Langevin Equations— Some Identities

This section presents several mathematical identities involving generalized Langevin equations.

Non-Markovian Fluctuation-Dissipation Theorem

One consequence of the general theory just presented is a non-Markovian version of the fluctuation-dissipation theorem. When the inner product is chosen to be the equilibrium average, we may rewrite the memory kernel in the form:

$$\mathbf{K}(t) = (F(t), LA) \cdot (A, A)^{-1} = (F(t), F(0)) \cdot (A, A)^{-1}, \tag{8.58}$$

where we have used the anti-Hermitian property of L, and we have inserted a redundant factor $(1 - \mathbf{P})$ to get $F(0) = (1 - \mathbf{P})LA$. This equation is in fact a generalization of the fluctuation-dissipation theorem,

$$\langle F(t) F^*(t') \rangle = \mathbf{K}(t - t') \cdot \langle A A^* \rangle. \tag{8.59}$$

This is like the result found in section 1.6, dealing with a harmonic oscillator heat bath. This is a mathematical identity, holding even if the initial distribution is not close to equilibrium.

Time Correlation Functions

Another useful consequence of this mathematical treatment is that equilibrium time correlation functions of the variables $\{A\}$ satisfy similar equations of motion, but without the noise term. The time correlation function is defined by the inner product of $A(t)$ and $A(0)$,

$$\mathbf{C}(t) = (A(t), A(0)). \tag{8.60}$$

If the inner product is the equilibrium average, then $\mathbf{C}(t)$ is the usual equilibrium time correlation function. To get the equation of motion for this matrix, we start with the Langevin equation and take its inner product with the initial $A(0)$, which is just the variable A. The noise term leads to $(F(t), A)$, which vanishes because F is orthogonal to A. The result is

$$\frac{\partial}{\partial t} \mathbf{C}(t) = i\Omega \cdot \mathbf{C}(t) - \int_0^t ds\, \mathbf{K}(s) \cdot \mathbf{C}(t-s). \tag{8.61}$$

Equilibrium time correlation functions satisfy exact linear transport equations. This provides a powerful method for finding explicit expressions for $i\Omega$ and $\mathbf{K}$; one works backwards from information about the time correlation function. For example, $i\Omega$ is the initial time derivative of $\mathbf{C}$.

It is worth noting, in this connection, the special connection of $\mathbf{C}(t)$ to the relaxation of $\langle A; t\rangle$,

$$\langle A; t\rangle = \mathbf{C}(t) \cdot \mathbf{C}(0)^{-1} \cdot \langle A; 0\rangle, \tag{8.62}$$

which we found in section 8.2. This was a consequence of the same special choice of initial state that we used in deriving the Langevin equation.

The generalized fluctuation-dissipation theorem is exact for any inner product in which L is anti-Hermitian, and the equations for equilibrium time correlation functions defined as inner products are exact for any choice of inner product.

Eliminating the Projection

The memory kernel K in the generalized Langevin equation involves projected dynamics, with the operator $\exp(1-\mathbf{P})Lt$. This means that its calculation is likely to be hard. The response of an equilibrium system to an imposed external field does not require any projection operator, and its calculation, involving conventional dynamics, is likely to be easier. There is a simple relation between these two theories, which

PROJECTION OPERATORS

allows us to write the memory kernel in a way that does not contain projection operators.

The linear response of a system to the perturbation $H' = -A\, E(t)$ is given by

$$\langle A;t\rangle = \int_0^t ds\, \phi(s) E(t-s) + \cdots, \tag{8.63}$$

and the response function is connected simply with the equilibrium time correlation function of A, $\mathbf{C}(t) = \langle A(t)\, A\rangle$,

$$\begin{aligned}\phi(t) &= \beta\langle A(t)\dot{A}(0)\rangle \\ &= -\beta\langle LA(t) A\rangle \\ &= -\beta\frac{d}{dt}\langle A(t)A\rangle \\ &= -\beta\dot{C}(t).\end{aligned} \tag{8.64}$$

The Laplace transform of this equation is

$$\hat{\phi} = -\beta(z\hat{C} - C(0)). \tag{8.65}$$

But the generalized Langevin equation provides an expression for the transform of the same time correlation function,

$$\hat{C} = \frac{1}{z - i\Omega + \hat{K}} \cdot C(0). \tag{8.66}$$

By combining the last two equations, we find the connection

$$\hat{\phi} = -\beta \frac{1}{z - i\Omega + \hat{K}} \cdot (i\Omega - \hat{K}) \cdot \langle AA\rangle. \tag{8.67}$$

This can be solved for the memory kernel,

$$\hat{K} = -z + i\Omega - \beta z\langle AA\rangle \cdot \frac{1}{\hat{\phi} - \beta\langle AA\rangle}. \tag{8.68}$$

The memory function in the generalized Langevin equation is connected to the response to an external field. This connection can also be used to treat the combined response to an initially nonequilibrium state and to an external field.

Combined Responses

The treatment of linear response to an external field supposed that the system was initially in equilibrium; the treatment of Langevin equations supposed that there was no time-dependent external field. These two treatments can be combined easily by noting that the effects of an initial deviation from equilibrium and an imposed field are additive. By taking Laplace transforms, we can write

$$\langle \hat{A} \rangle = \frac{1}{z - i\Omega + \hat{K}} \langle A; 0 \rangle + \hat{\phi} \hat{E}. \tag{8.69}$$

The first term is the response to an initial deviation from equilibrium; the second term is the response to an imposed field. Equation (8.68) provides an expression we can use to eliminate the response function $\hat{\phi}$, and so by multiplying out, the average obeys

$$(z - i\Omega + \hat{K}) \cdot \langle \hat{A} \rangle = \langle A; 0 \rangle - \beta (i\Omega - \hat{K}) \cdot \langle AA \rangle \cdot \hat{E}. \tag{8.70}$$

The inverse Laplace transform of this equation, its time-dependent version, is

$$\frac{d}{dt} \langle A; t \rangle = i\Omega \cdot \delta \langle A; t \rangle - \int_0^t ds\, K(t-s) \cdot \delta \langle A; t \rangle. \tag{8.71}$$

Note that the right-hand side contains the quantity

$$\delta \langle A; t \rangle = \langle A; t \rangle - \beta \langle AA \rangle \cdot E(t), \tag{8.72}$$

which is the deviation of the average from what it would be if the system were in local equilibrium at time t. A special case, arising in the treatment of electrolyte solutions, is known as the Nernst-Planck equation.

8.5 From Nonlinear to Linear—An Example

Earlier, we noted a subtle difference between Mori's linear generalized Langevin equation and the exact Langevin equation for a nonlinear system interacting with a harmonic oscillator heat bath. ("Nonlinear" means here that the potential energy contains higher powers than quadratic.) We now use a special example to see how these two levels of description are related. This example is simple enough to allow an explicit calculation of the noise and memory function in Mori's equation. We find that noise in the linear generalized Langevin equation comes not only from noise in the nonlinear Langevin equation but also from certain explicit effects of the nonlinearity. The same holds for the memory functions. This change in memory functions, when it occurs in connection with fluid transport coefficients, is sometimes called "fluctuation-renormalization." We return to this connection later in a discussion of the long time asymptotic decay of the velocity correlation function of a particle in a fluid.

In this special example, the system Hamiltonian is

$$H = \frac{1}{2}v^2 + \frac{1}{2}x^2 + \frac{b}{4}x^4 + \sum_j \left\{ \frac{1}{2}p_j^2 + \frac{1}{2}\omega_j^2 \left(q_j - \frac{\gamma_j}{\omega_j^2}x\right)^2 \right\}. \tag{8.73}$$

The parameter b measures the strength of the nonlinearity; we look for the Mori memory function in the limit of small b. For convenience, the system mass is set equal to 1. The heat bath is the same one used in section 1.6. The *exact* nonlinear Langevin equation for this model was derived there:

$$\frac{dx(t)}{dt} = v(t)$$

$$\frac{dv(t)}{dt} = -x(t) - bx(t)^3 - \int_0^t ds K_N(s)v(t-s) + F_N(t), \tag{8.74}$$

where the memory function and noise are labeled by a subscript N to indicate that they are appropriate to the nonlinear problem,

$$K_N(t) = \sum_j \frac{\gamma_j^2}{\omega_j^2} \cos\omega_j t$$

$$F_N(t) = \sum_j {}_j\gamma_j p_j(0) \frac{\sin\omega_j t}{\omega_j} + \sum_j {}_j\gamma_j \left(q_j(0) - \frac{\gamma_j}{\omega_j^2} x(0) \right) \cos\omega_j t. \tag{8.75}$$

The correlation function of the noise is the generalized fluctuation-dissipation theorem,

$$\langle F_N(t) F_N(t') \rangle^0 = k_B T K_N(t-t'). \tag{8.76}$$

The average is over a constrained equilibrium heat bath.

We can apply Mori's procedure to the same Hamiltonian, by projecting onto the subspace of the variables x and v. The projection operator is explicitly

$$PB = \langle Bx \rangle \langle xx \rangle^{-1} x + \langle Bv \rangle \langle vv \rangle^{-1} v. \tag{8.77}$$

The equilibrium second moments are

$$\langle v^2 \rangle = k_B T, \quad \langle x^2 \rangle = \frac{k_B T}{\omega_0^2}, \tag{8.78}$$

which defines the frequency ω_0. Then Mori's procedure leads to the *exact* linear equations of motion

$$\frac{dx(t)}{dt} = v(t)$$
$$\frac{dv(t)}{dt} = -\omega_0^2 x(t) - \int_0^t ds K_L(s) v(t-s) + F_L(t). \quad (8.79)$$

The memory function and noise are distinguished here by the subscript L to indicate that they are appropriate to the linear Langevin equation. The noise is given explicitly by the operator expression,

$$F_L(t) = e^{t(1-P)L}(1-P)Lv, \quad (8.80)$$

and the memory function is the equilibrium correlation function of $F_L(t)$,

$$\langle F_L(t) F_L(t') \rangle = k_B T K_L(t-t'). \quad (8.81)$$

The average is over the unconstrained thermal equilibrium distribution.

It is easy to work out the relationship of F_L to F_N in the limit of small b. The frequency ω_0 to first order in b is

$$\omega_0^2 = 1 + 3k_B T b + O(b^2). \quad (8.82)$$

The complete Liouville operator may be separated into a linear part L_0 and a perturbation L_1,

$$L = L_0 + L_1, \quad L_1 = -bx^3 \frac{\partial}{\partial v}. \quad (8.83)$$

Then $(1-P)Lv$ separates into two parts,

$$(1-P)Lv = \sum_j \gamma_j \left(q_j - \frac{\gamma_j}{\omega_j^2} x \right) + [(\omega_0^2 - 1)x - bx^3]. \quad (8.84)$$

To first order in b, the second term is

$$(\omega_0^2 - 1)x - bx^3 = b(3k_B Tx - x^3) + O(b^2). \quad (8.85)$$

The projected time evolution operator may also be expanded to first order,

$$e^{t(1-P)L} = e^{t(1-P)L_0} + \int_0^t ds \, e^{(t-s)(1-P)L_0}(1-P)L_1 e^{s(1-P)L_0} + \cdots. \quad (8.86)$$

To first order, the total force F_L consists of three terms,

PROJECTION OPERATORS 163

$$F_L(t) = e^{t(1-P)L_0} \sum \gamma_j \left(q_j - \frac{\gamma_j}{\omega_j^2} x \right) + e^{t(1-P)L_0} [(\omega_0^2 - 1)x - bx^3]$$

$$+ \int_0^t ds\, e^{(t-s)(1-P)L_0}(1-P)L_1 e^{s(1-P)L_0} \sum \gamma_j \left(q_j - \frac{\gamma_j}{\omega_j^2} x \right). \quad (8.87)$$

The first term, denoted by $F_0(t)$, is easy to evaluate. When the operator L_0 operates on any linear combination of variables $\{x, v, p_j, q_j\}$, it produces a new linear combination of these variables. Further, the projection $(1 - P)$ maintains this linearity. Thus the projected time evolution operator converts a linear combination into a new linear combination. This suggests the structure

$$F_0(t) = \rho(t)x + \sigma(t)v + \sum \mu_j(t) q_j + \sum v_j(t) p_j, \quad (8.88)$$

where the coefficients $\{\sigma, \rho, \mu_j, v_j\}$ are functions of time only. F_0 obeys the operator equation

$$\frac{\partial}{\partial t} F_0(t) = (1 - P) L_0 F_0(t). \quad (8.89)$$

On substituting the assumed form of F_0, the effect of the projection is contained in

$$P L_0 F_0(t) = \rho(t)v - \sigma(t)x. \quad (8.90)$$

When coefficients of the dynamical variables are collected, one finds that the coefficients obey the linear equations

$$\dot{\sigma}(t) = 0$$
$$\dot{\rho}(t) = -\sigma(t) \sum \frac{\gamma_j^2}{\omega_j^2} + \sum v_j(t) \gamma_j$$
$$\dot{v}_j(t) = \mu_j(t)$$
$$\dot{\mu}_j(t) = \sigma(t) \gamma_j - \omega_j^2 v_j(t). \quad (8.91)$$

Furthermore, the initial values of the coefficients are

$$\sigma(0) = v_j(0) = 0$$
$$\mu_j(0) = \gamma_j; \quad \rho(0) = -\sum \frac{\gamma_j^2}{\omega_j^2}. \quad (8.92)$$

Thus $\sigma(t)$ vanishes for all t. The equations for μ_j and v_j are ordinary harmonic oscillator equations that can be easily solved:

$$\mu_j(t) = \cos(\omega_j t)\gamma_j, \quad v_j(t) = \frac{\sin(\omega_j t)}{\omega_j}\gamma_j, \tag{8.93}$$

and finally, the equation for $\rho(t)$ can be integrated to give

$$\rho(t) = -\sum \frac{\gamma_j^2}{\omega_j^2}\cos(\omega_j t). \tag{8.94}$$

The result is an expression for $F_0(t)$,

$$F_0(t) = \sum \gamma_j p_j \frac{\sin(\omega_j t)}{\omega_j} + \sum \gamma_j \left(q_j - \frac{\gamma_j}{\omega_j^2}x\right)\cos(\omega_j t), \tag{8.95}$$

which is clearly identical to the earlier $F_N(t)$.

The third term in $F_L(t)$ vanishes. It can be written as

$$\int_0^t ds\, e^{(t-s)(1-P)L_0}(1-P)L_1 F_0(s). \tag{8.96}$$

But F_0 is independent of v, and $L_1 F_0(s)$ vanishes, so the entire term vanishes.

The second term in F_L requires further attention:

$$F_1(t) = be^{t(1-P)L_0}(3k_B T x - x^3) + O(b^2). \tag{8.97}$$

On taking the time derivative, one finds

$$\frac{\partial}{\partial t}F_1 = (1-P)L_0 F_1. \tag{8.98}$$

But the term with the projection can be dropped, $PL_0 F_1 = 0$, because the two inner products $(L_0 F_1, x)$ and $(L_0 F_1, v)$ both vanish. Thus $F_1(t)$ is determined by the unperturbed (i.e., linear) motion of the system and bath,

$$F_1(t) = be^{tL_0}(3k_B T x - x^3) + O(b^2). \tag{8.99}$$

The total random force in the linear Langevin equation, to first order in the nonlinearity parameter b, is

$$F_L(t) = F_N(t) + be^{tL_0}(3k_B T x - x^3) + O(b^2). \tag{8.100}$$

This is evidently not the same force that appears in the exact nonlinear Langevin equation.

Because F_N falls in the linear subspace and F_1 falls in the complementary subspace, they are orthogonal for all times. This means that the memory function $K_L(t)$ separates into two parts,

$$K_L(t) = K_N(t)$$
$$+ (b^2/k_BT)\langle(3k_BTx - x^3)e^{tL_0}(3k_BTx - x^3)\rangle + O(b^3). \tag{8.101}$$

(Recall that L_0 is the complete Liouville operator for the linearized system, $b = 0$.)

This example makes it clear that different kinds of noise can appear in a Langevin equation, depending on the level of description. The noise in the nonlinear Langevin equation appears to be more "intrinsic"; in particular, by making an appropriate choice of heat bath parameters, F_N can be approximated by white noise. However, its statistical properties are simple only for a particular class of initial distributions. The noise F_L in the Mori Langevin equation, which is what one would see in a study of fluctuations in an equilibrium system, differs from F_N because of nonlinear effects. The Mori memory function $K_L(t)$ will have a Markovian part, connected with the intrinsic white noise, but also a non-Markovian part coming from the nonlinearity in the system Hamiltonian. This turns out to be a common situation.

8.6 Linear Langevin Equations for Slow Variables

Many dynamical variables of practical interest are "slow"; their rates of change are controlled by a small parameter λ,

$$\frac{\partial}{\partial t} A(t) = LA(t) = 0(\lambda). \tag{8.102}$$

One example has already appeared, in the derivation of the master equation. Here we discuss linear Langevin equations for slow variables.

In the generalized Langevin equation, the quantity $i\Omega$, containing one factor LA, is of order λ, and the memory kernel K, containing two factors of LA, is formally of order λ^2. In the limit of small λ, the time convolution of the generalized Langevin equation can be replaced by its Markovian approximation,

$$\frac{d}{dt} A(t) \cong i\Omega \cdot A(t) - \int_0^\infty ds K(s) \cdot A(t) + F(t) + O(\lambda^3 A). \tag{8.103}$$

But the memory kernel still contains the exponential of the projected Liouville operator $(1 - \mathbf{P})L$, and this is hard to work with. Fortunately, however, the projection operator can be dropped when one deals with slow variables. The argument is as follows. We start with the identity

$$e^{tL} = e^{t(1-\mathbf{P})L} + \int_0^t ds e^{(t-s)L} \mathbf{P} L e^{s(1-\mathbf{P})L}. \tag{8.104}$$

Then we recall that, from the definition of the projection operator,

$$\mathbf{P}LB = (LB, A) \cdot (A, A)^{-1} \cdot A = -(B, LA) \cdot (A, A)^{-1} \cdot A, \tag{8.105}$$

and this quantity, containing one factor of LA, is of order λ. Consequently, the difference between projected and ordinary dynamics is of the same order,

$$e^{(1-\mathbf{P})Lt} = e^{Lt} + O(\lambda). \tag{8.106}$$

The general formula for $\mathbf{K}(t)$ is

$$\mathbf{K}(t) = (e^{(1-\mathbf{P})Lt}(1-\mathbf{P})LA, LA) \cdot (A, A)^{-1}. \tag{8.107}$$

Because the factors on the left side of the inner product all start with $(1 - \mathbf{P})$, we can insert a redundant $(1 - \mathbf{P})$ in front of the LA on the right side,

$$\mathbf{K}(t) = (e^{(1-\mathbf{P})Lt}(1-\mathbf{P})LA, (1-\mathbf{P})LA) \cdot (A, A)^{-1}. \tag{8.108}$$

Finally, we use eq. (8.106) to simplify the exponential operator,

$$\mathbf{K}(t) = (e^{Lt}(1-\mathbf{P})LA, (1-\mathbf{P})LA) \cdot (A, A)^{-1} + O(\lambda^3). \tag{8.109}$$

To second order in λ, the memory kernel involves the conventional (i.e., unprojected) time correlation function of the quantity $(1 - \mathbf{P})LA$ or $LA - i\Omega \cdot A$.

Then the Mori linear Langevin equation for a slow variable, or a set of slow variables, has the standard form

$$\frac{\partial}{\partial t} A \cong \Theta \cdot A + F(t). \tag{8.110}$$

In particular, the hydrodynamic variables, mass density, momentum density, and energy density, and in a multicomponent mixture, composition, are all slow variables.

Self-Diffusion

A first example is self-diffusion. Here the dynamical quantity of interest is the concentration $C(x, t)$ of the tagged particle at x and t. (We assume that the concentration depends on only one coordinate.) This has the spatial Fourier transform,

$$C(x, t) = \sum_q A_q(t) e^{iqx}. \tag{8.111}$$

PROJECTION OPERATORS 167

The Fourier component is a function of the position R of the tagged particle,

$$A_q = e^{-iqR}. \tag{8.112}$$

Its rate of change is

$$LA_q = -iq\dot{R}e^{-iqR} = -iqV + O(q^2), \tag{8.113}$$

where V is the velocity of the tagged particle. Because $i\Omega$ is the average of an odd function of V, it vanishes. The memory function to second order is

$$K(t) = -q^2 \langle Ve^{tL}V \rangle + O(q^3). \tag{8.114}$$

It contains the familiar velocity correlation function $\langle V(t)V \rangle$. The average concentration obeys

$$\frac{d}{dt}C_q(t) = -q^2 \int_0^t ds \langle V(s)V \rangle C_q(t-s) + O(q^3). \tag{8.115}$$

In the long wavelength or small q limit, A_q is a slow variable, and we can make the Markovian approximation

$$\frac{d}{dt}A_q(t) \cong -q^2 \int_0^\infty ds \langle V(s)V \rangle A_q(t) + O(q^3). \tag{8.116}$$

This is the Fourier transform of the diffusion equation

$$\frac{d}{dt}C(x,t) \cong D\nabla^2 C(x,t), \tag{8.117}$$

and the self-diffusion coefficient is the familiar time integral of the VCF.

Hydrodynamics

Hydrodynamic variables, the mass density, momentum density, and energy density of a fluid, are also slow. In an N-body system, particles are labeled $j = 1, 2, 3, \ldots$ and are located at positions $\mathbf{R}_j$. The jth particle has mass m_j, momentum $\mathbf{p}_j$, and energy ε_j. Spatial Fourier components of the mass, momentum, and energy densities are labeled by the vector $\mathbf{q}$,

$$\rho_\mathbf{q} = \sum_j m_j e^{i\mathbf{q}\cdot\mathbf{R}_j}, \quad \mathbf{J}_\mathbf{q} = \sum_j \mathbf{p}_j e^{i\mathbf{q}\cdot\mathbf{R}_j}, \quad E_\mathbf{q} = \sum_j \varepsilon_j e^{i\mathbf{q}\cdot\mathbf{R}_j}. \tag{8.118}$$

The time derivative of any of these sums can be expanded in powers of **q**,

$$L\sum_j A_j e^{i\mathbf{q}\cdot\mathbf{R}_j} = L\sum_j A_j + L\sum_j i\mathbf{q}\cdot\mathbf{R}_j A_j + 0(\mathbf{q}^2)$$
$$= i\mathbf{q}\cdot\sum_j L(\mathbf{R}_j A_j) + 0(\mathbf{q}^2). \qquad (8.119)$$

The zeroth order term vanishes because the total mass, total momentum, and total energy are conserved quantities, invariant to the Liouville operator. This leaves the first-order term. The magnitude of the Fourier vector **q** is the small parameter required for a "slow" variable. If **q** is small enough, the hydrodynamic variables are slow.

A particular application is to shear flow, where the fluid velocity $v_x(y,t)$ points in the x direction and varies spatially only in the y direction. This particular velocity field satisfies a special case of the Navier-Stokes equation,

$$\rho\frac{\partial}{\partial t}v_x = \eta\frac{\partial^2}{\partial y^2}v_x. \qquad (8.120)$$

The Fourier expansion of the momentum density ρv_x is

$$(\rho v_x)_q = \sum_j p_{jx} e^{iqy_j}, \qquad (8.121)$$

so the rate of change in eq. (8.119) is

$$L\sum_j p_{jx} y_j = \sum_j \left(\frac{p_{jx}p_{jy}}{m} + F_{jx}y_j\right) = \mathbf{P}_{xy}. \qquad (8.122)$$

This is the xy component of the molecular stress tensor **P**; its time correlation function determines the shear viscosity η,

$$\eta = \frac{1}{VkT}\int_0^\infty dt\langle\mathbf{P}_{xy}(t)\mathbf{P}_{xy}(0)\rangle. \qquad (8.123)$$

A Warning

As was seen earlier, in the discussion of quantum mechanical models where a two-state system interacts weakly with a heat bath, one must be careful about deciding what is to be treated as "slow." It may happen that even though the memory function **K** is small, the $i\Omega$ coefficient can still be large. This has to be taken into account when making a Markovian approximation. Fortunately, applications to self-diffusion and hydrodynamics do not face this difficulty because there $i\Omega$ is already small, of order **q**.

9

Nonlinear Problems

9.1 Mode-Coupling Theory and Long Time Tails

This chapter deals with nonlinear Langevin equations and related topics. The first topic is mode-coupling theory.

Mode-coupling theory refers to a family of techniques that were invented to explain anomalous properties of certain transport coefficients. This chapter presents one particular version of the theory, due to L. Kadanoff and J. Swift (1968). Another version is discussed in the following sections.

The anomalous properties referred to occur in two quite different contexts. One is critical phenomena. For example, the thermal conductivity of a fluid shows singular behavior in the neighborhood of the critical point of the liquid-gas phase transition. Because a detailed treatment of such behavior requires a thorough knowledge of the equilibrium theory of critical phenomena, which is beyond the scope of this book, we will not go further into critical dynamics.

The other context is "long time tails." B. Alder and T. Wainwright (1968) made an extensive computer study of the molecular dynamics of a hard disk fluid (d = two dimensions) and a hard sphere fluid (d = three dimensions). They observed that the velocity correlation function of a tagged particle in the fluid decays for a long time as an inverse power of the time,

$$D(t) = \frac{1}{d}\langle \mathbf{V}(t) \cdot \mathbf{V} \rangle \propto t^{-d/2}, \qquad d = 2, 3. \tag{9.1}$$

170 NONEQUILIBRIUM STATISTICAL MECHANICS

The specific power is determined only by the dimensionality of the system. Later, similar behavior was seen for the stress time correlation function that leads to the fluid viscosity. This kind of asymptotic behavior is called a "long time tail."

Mode-coupling theory is ultimately based on the Hilbert space picture of dynamics that was discussed in section 8.1. This starts with a complete orthonormal set of functions $\varphi_j(X)$ of the position X of the system in phase space, with the inner product

$$(\varphi_j, \varphi_k) = \int dX \varphi_j(X) \varphi_k^*(X) f_{eq}(X) = \delta_{jk}. \tag{9.2}$$

Then any time correlation function, in particular the VCF, can be represented by the expansion

$$D = \int_0^\infty dt\, D(t)$$

$$D(t) = \frac{1}{d}\langle \mathbf{V}(t)\cdot \mathbf{V}\rangle = \frac{1}{d}\mathrm{Trace}\sum_{j,k}(\mathbf{V}, \varphi_j)(e^{tL}\varphi_j, \varphi_k)(\varphi_k, \mathbf{V}). \tag{9.3}$$

Clearly this shifts the dynamical problem from finding $\mathbf{V}(t)$ into finding $e^{tL}\varphi_j$. In general, this is a hard problem, but for some special functions φ_j it is easy. These are the functions that can be constructed from slow variables.

In the present example, self-diffusion, one slow variable is the concentration $C(\mathbf{r}, t) = \delta(\mathbf{R}_0(t) - \mathbf{r})$ of the tagged particle at position $\mathbf{r}$. This (approximately) satisfies a diffusion equation with noise,

$$\frac{\partial}{\partial t}C(\mathbf{r}, t) = D\nabla^2 C(\mathbf{r}, t) + \text{noise}. \tag{9.4}$$

On taking a spatial Fourier transform,

$$C(\mathbf{r}, t) = \sum_q C_q(t) e^{i\mathbf{q}\cdot\mathbf{r}}, \tag{9.5}$$

the diffusion equation becomes

$$\frac{\partial}{\partial t}C_q = -Dq^2 C_q + \text{noise}. \tag{9.6}$$

R_0 is the position of the tagged particle; then

$$C_q = e^{-i\mathbf{q}\cdot\mathbf{R}_0}, \tag{9.7}$$

and if the noise can be ignored, $C_q(t)$ is approximately

$$C_{\mathbf{q}}(t) = e^{tL}C_{\mathbf{q}} \cong e^{-Dq^2 t}C_{\mathbf{q}}. \tag{9.8}$$

When self-diffusion takes place in a fluid, diffusive transport can be modified by convective transport due to fluctuations in the fluid velocity $\mathbf{v}(\mathbf{r}, t)$. But hydrodynamics tells us that the fluid velocity is another slow variable. The explicit form of its spatial Fourier transform, $\mathbf{v}_{\mathbf{q}}$, is

$$\mathbf{v}_{\mathbf{q}} = \sum_j \frac{\mathbf{p}_j}{m} e^{-i\mathbf{q}\cdot\mathbf{R}_j}, \tag{9.9}$$

where $\mathbf{p}_j$ and $\mathbf{R}_j$ are the momentum and position of the jth particle. One of these particles, $j = 0$, is the tagged one, and $\mathbf{V} = \mathbf{p}_0/m$.

Now we return to the Hilbert space expansion of the VCF. Most of the functions φ_j are unrelated to the slow variables and may be called "fast." These give a contribution to the VCF that decays rapidly. However, some of the φ_j can be constructed from slow variables and give a slowly decaying contribution to the VCF, a long time tail. An obvious first choice for φ_j is one of the variables $C_{\mathbf{q}}$ or $\mathbf{v}_{\mathbf{q}}$; these may be called "single mode" functions. However, they cannot contribute to the VCF; the inner products vanish,

$$(\mathbf{V}, C_{\mathbf{q}}) = 0, \qquad (\mathbf{V}, \mathbf{v}_{\mathbf{q}}) = 0 \tag{9.10}$$

because $\mathbf{V}$ is independent of position, and if $\mathbf{q} \neq 0$, $C_{\mathbf{q}}$ and $\mathbf{v}_{\mathbf{q}}$ are not.

The next simplest choice for φ_j is a product of two slow variables or modes, hence the name "mode-coupling." So that the inner product with $\mathbf{V}$ does not vanish, this product must contain one factor of $\mathbf{v}_{\mathbf{q}}$ to provide a velocity to match $\mathbf{V}$, and translational symmetry requires that this be accompanied by one factor of $C_{-\mathbf{q}}$. Then this inner product is

$$(\mathbf{V}, \mathbf{v}_{\mathbf{q}} C_{-\mathbf{q}}) = \left\langle \mathbf{V}(\mathbf{v}_{\mathbf{q}} C_{-\mathbf{q}})^* \right\rangle$$

$$= \left\langle \mathbf{V} \sum_j \frac{\mathbf{p}_j}{m} e^{i\mathbf{q}\cdot\mathbf{R}_j} e^{-i\mathbf{q}\cdot\mathbf{R}_0} \right\rangle = \left\langle \mathbf{V} \frac{\mathbf{p}_0}{m} e^{i\mathbf{q}\cdot\mathbf{R}_0} e^{-i\mathbf{q}\cdot\mathbf{R}_0} \right\rangle = \frac{kT}{m} \mathbf{1}. \tag{9.11}$$

(Recall that $\mathbf{V} = \mathbf{p}_0/m$.) For normalization, we also need the inner product

$$(\mathbf{v}_{\mathbf{q}} C_{-\mathbf{q}}, \mathbf{v}_{\mathbf{q}} C_{-\mathbf{q}}) = \langle \mathbf{v}_{\mathbf{q}} \mathbf{v}_{-\mathbf{q}} \rangle \langle C_{-\mathbf{q}} C_{\mathbf{q}} \rangle = \frac{NkT}{m} \mathbf{1}. \tag{9.12}$$

Then, on replacing the index m by the wave vector q, the normalized expansion function is

$$\varphi_{\mathbf{q}} = (m/NkT)^{1/2} \mathbf{v}_{\mathbf{q}} C_{-\mathbf{q}}. \tag{9.13}$$

As a result of translational invariance, these functions are orthogonal,

$$(\varphi_{\mathbf{q}}, \varphi_{\mathbf{q}'}) = \delta_{\mathbf{q},\mathbf{q}'}. \tag{9.14}$$

Further, because the time evolution of both $\mathbf{v}_{\mathbf{q}}$ and $C_{-\mathbf{q}}$ is given by linear equations with coefficients that depend on $\mathbf{q}$, their time correlation is orthogonal,

$$(e^{tL}\varphi_{\mathbf{q}}, \varphi_{\mathbf{q}'}) = (e^{tL}\varphi_{\mathbf{q}}, \varphi_{\mathbf{q}})\delta_{\mathbf{q},\mathbf{q}'}. \tag{9.15}$$

Then the VCF is

$$D(t) = \frac{1}{d}D_{\text{fast}}(t) + \frac{1}{d}\text{trace}\sum_{\mathbf{q}}(\mathbf{V}, \varphi_{\mathbf{q}})(e^{tL}\varphi_{\mathbf{q}}, \varphi_{\mathbf{q}})(\varphi_{\mathbf{q}}, \mathbf{V}) + \cdots$$

$$= \frac{1}{d}D_{\text{fast}}(t) + \frac{1}{d}\text{trace}\sum_{\mathbf{q}}(e^{tL}\mathbf{v}_{\mathbf{q}}C_{-\mathbf{q}}, \mathbf{v}_{\mathbf{q}}C_{-\mathbf{q}}) + \cdots, \tag{9.16}$$

where $+\ldots$ refers to the contribution from neglected higher order mode-coupling products. Because the Liouville operator can be distributed over products, the effect of e^{tL} is

$$e^{tL}\mathbf{v}_{\mathbf{q}}C_{-\mathbf{q}} = (e^{tL}\mathbf{v}_{\mathbf{q}})(e^{tL}C_{-\mathbf{q}}) \cong (e^{tL}\mathbf{v}_{\mathbf{q}})e^{-Dq^2t}C_{-\mathbf{q}}. \tag{9.17}$$

The fluid velocity is a vector depending on $\mathbf{q}$ and can be decomposed into longitudinal (ℓ) and transverse ($\perp$) parts,

$$\mathbf{v}_{\mathbf{q}} = \mathbf{v}_{\mathbf{q}\ell} + \mathbf{v}_{\mathbf{q}\perp}, \quad \mathbf{v}_{\mathbf{q}\ell} = \frac{\mathbf{qq}}{q^2}\cdot\mathbf{v}_{\mathbf{q}}, \quad \mathbf{v}_{\mathbf{q}\perp} = \left(1 - \frac{\mathbf{qq}}{q^2}\right)\cdot\mathbf{v}_{\mathbf{q}}. \tag{9.18}$$

To make the discussion simpler, we assume that the fluid is incompressible and approximately satisfies the Navier-Stokes equation,

$$\rho\frac{\partial}{\partial t}\mathbf{v} = -\nabla P + \eta\nabla^2\mathbf{v} + \text{noise}, \quad \nabla\cdot\mathbf{v} = 0. \tag{9.19}$$

Then the transverse Fourier components satisfy

$$\frac{\partial}{\partial t}\mathbf{v}_{\mathbf{q}\perp} = -\nu q^2\mathbf{v}_{\mathbf{q}\perp} + \text{noise}. \tag{9.20}$$

Here we abbreviate $\eta/\rho = \nu$, the kinematic viscosity. If the noise is neglected, the time evolution of $\mathbf{v}_{\mathbf{q}\perp}$ is given approximately by

$$\mathbf{v}_{\mathbf{q}\perp}(t) = e^{tL}\mathbf{v}_{\mathbf{q}\perp} \cong e^{-\nu q^2 t}\mathbf{v}_{\mathbf{q}\perp}. \tag{9.21}$$

NONLINEAR PROBLEMS   173

In an incompressible fluid, there is no longitudinal part. The TCF of the transverse part is

$$(e^{tL}\mathbf{v}_{\mathbf{q}\perp}, \mathbf{v}_{\mathbf{q}\perp}) \cong e^{-vq^2t}(\mathbf{v}_{\mathbf{q}\perp}, \mathbf{v}_{\mathbf{q}\perp}) = \frac{NkT}{m}\left(1 - \frac{\mathbf{qq}}{q^2}\right)e^{-vq^2t}. \tag{9.22}$$

The resulting VCF is

$$D(t) = D_{\text{fast}}(t) + \frac{1}{d}\text{Trace}\frac{kT}{mN}\sum_{\mathbf{q}}e^{-(D+v)q^2t}\left(1 - \frac{\mathbf{qq}}{q^2}\right) + \cdots$$

$$= D_{\text{fast}}(t) + \frac{d-1}{d}\frac{kT}{mN}\sum_{\mathbf{q}}e^{-(D+v)q^2t} + \cdots. \tag{9.23}$$

In the limit of a large system, the sum over $\mathbf{q}$ can be replaced by an integral,

$$\sum_{\mathbf{q}} \to \left(\frac{L}{2\pi}\right)^d \int d^dq, \tag{9.24}$$

where L^d is the area or volume of the system. The mass density is

$$\rho = \frac{mN}{L^d}, \tag{9.25}$$

and the q integral leads to

$$D(t) = D_{\text{fast}}(t) + \frac{d-1}{d}\frac{kT}{\rho}\frac{1}{[4\pi(D+v)t]^{d/2}} + \cdots. \tag{9.26}$$

Note that this result is not correct for short times. The integration over q should be terminated at large q when $1/q$ is of the order of a molecular size; this cut-off removes the singular behavior at small t but does not affect the long time behavior.

The $1/t$ decay at long times for $d = 2$ means that in this approximation, the self-diffusion coefficient in a two-dimensional fluid, which is the integral of the VCF over all time, does not exist. This is consistent with Stokes' paradox, that the hydrodynamic friction on a particle moving in a fluid does not exist in two dimensions.

Another use of this process is to derive the long time tail of the stress correlation function that determines the viscosity of a fluid. As seen in the last section, the stress tensor is quadratic in particle velocity. Then it couples naturally with a product of two fluid velocity modes, $\mathbf{v}_{\mathbf{q}}\mathbf{v}_{-\mathbf{q}}$. The result is that the stress correlation function has a long time tail $t^{-3/2}$ in three dimensions.

174   NONEQUILIBRIUM STATISTICAL MECHANICS

Clearly the success of this method depends critically on an intelligent choice of product modes. One should always check to see whether more-complex mode-couplings are likely to contribute significantly, but this is usually not easy to do. A method based on nonlinear Langevin equations for slow variables, described in a later section, provides a more systematic procedure.

Mode-coupling theory leads to a separation of time scales—part of a time correlation function decays very rapidly and part decays slowly. The transport coefficient determined by this TCF is non-Markovian at long times. For example, the non-Markovian diffusion coefficient now depends on a Laplace transform variable z,

$$\hat{D}(z) = \int_0^\infty dt\, e^{-zt} D(t)$$

$$= D_{\text{fast}} + \frac{d-1}{d}\frac{kT}{\rho}\frac{1}{(2\pi)^d}\int d^d q \frac{1}{z+(D+v)q^2} + \cdots. \qquad (9.27)$$

(The q integral must be cut off when $1/q$ reaches a molecular size.) The denominator in the integral contains the "bare" diffusion coefficient D, and the resulting expression is a non-Markovian or "renormalized" $\hat{D}$. It is often suggested that better results can be obtained if one replaces the "bare" D in the denominator by $\hat{D}$. This leads to a self-consistent mode-coupling equation to be solved for $\hat{D}$,

$$\hat{D}(z) \approx D_{\text{fast}} + \frac{d-1}{d}\frac{kT}{\rho}\frac{1}{(2\pi)^d}\int d^d q \frac{1}{z+(\hat{D}+v)q^2}. \qquad (9.28)$$

This approximation generally has no real justification.

9.2 Derivation of Nonlinear Langevin and Fokker-Planck Equations

Introduction

The linear Langevin equations derived in earlier chapters, while formally exact, are useful only for systems near equilibrium. They inevitably lead to linear transport laws for averaged variables and cannot be used to treat nonlinear transport processes. It is possible to use the same procedure to derive nonlinear Langevin equations; the idea is to expand the projected space to include not only first powers of the dynamical variables $\{A_j\}$ but also higher powers, for example, product variables $\{A_j, A_k\}$. This is the basic idea of the mode-coupling theory that was discussed in the last section. An easier procedure (S. Nordholm, 1975) provides both Langevin and Fokker-Planck equations for nonlinear processes at the same time.

NONLINEAR PROBLEMS   175

Section 2.2 gave a conventional treatment of Fokker-Planck equations. We started there with a postulated Langevin equation for a set of variables **a**, with a possibly nonlinear streaming term **v(a)**,

$$\frac{\partial}{\partial t}\mathbf{a} = \mathbf{v}(\mathbf{a}) + \mathbf{F}(t); \tag{9.29}$$

we assumed that the noise **F**(t) was Gaussian and white, with the second moment

$$\langle \mathbf{F}(t)\mathbf{F}(t')\rangle = 2\mathbf{B}\delta(t-t'); \tag{9.30}$$

and we derived the resulting Fokker-Planck equation for the noise-averaged distribution function,

$$\frac{\partial}{\partial t}f = -\frac{\partial}{\partial \mathbf{a}}\cdot(\mathbf{v}(\mathbf{a})f) + \frac{\partial}{\partial \mathbf{a}}\cdot\mathbf{B}\cdot\frac{\partial}{\partial \mathbf{a}}f. \tag{9.31}$$

In this section, we derive equations of this form starting with the Liouville equation. As in the derivation of generalized Langevin equations, some of the results are formally exact, but structurally complex, with non-Markovian behavior in time and nonlocal behavior in **a**-space. As such, they are not pleasant to look at and not likely to have much practical utility. However, when the dynamical variables are "slow," as in section 8.6, the results of the derivation reduce to the simple form given above. Further, the results provide a straightforward statistical mechanical method for calculating the streaming velocity **v(a)** and the "diffusion coefficient" **B**. In this section, we consider only the formal derivation. The slow variable approximation is discussed in the following section.

Reduced Distribution Functions

The main technical problem in deriving Fokker-Planck equations from the full-phase space Liouville equation is that they describe the evolution of reduced distribution functions. In Brownian motion, for example, the reduced distribution function contains only the velocity V and position R of the Brownian particle. The full distribution function $f(X, t)$ depends on the entire set of phase space coordinates and momenta denoted by X. These variables can be separated into system variables (position and velocity of the Brownian particle) denoted by S, and heat bath variables denoted by B, so that $X = (S, B)$. The reduced distribution function $g(S, t)$ is obtained from the full distribution function by integrating out the bath variables,

$$g(S,t) = \int dB f(S, B, t). \tag{9.32}$$

(Note that we have switched notation. From here on, we use g when referring to reduced distribution functions and reserve f for the complete phase space function.)

Reduced distribution functions can also be constructed by using delta functions. In the present case, we can ask for the probability density that the system variables $S = (V, R)$ take on the numerical values $s = (v, r)$; then this can be written as the full phase space average of the product $\delta(R - r)\,\delta(V - v)$ of two delta functions, abbreviated by $\delta(S - s)$,

$$g(s,t) = \int dX\, \delta(S - s) f(X, t). \tag{9.33}$$

Often we want to deal with properties of a system other than a subset of its coordinates and momenta. Let us focus on a particular property $A(X)$ of the system. The probability that this variable has a numerical value in the interval between a and $a + da$ is denoted by $g(a, t)da$,

$$\text{prob}(a < A(X, t) < a + da) = g(a, t)da. \tag{9.34}$$

If this variable is not one of the canonical coordinates or momenta of the entire system, we can not get $g(a, t)$ by integrating over "bath" variables, but we can still use the delta function procedure. Then the reduced distribution function or probability density can be found from the phase space average,

$$g(a,t) = \int dX \delta(A(X) - a) f(X, t). \tag{9.35}$$

The delta function selects that part of phase space in which the dynamical variable A has the specified numerical value a. This "surface of constant A" is analogous to the surface of constant energy that is often used to define the equilibrium microcanonical ensemble. The phase space average of the delta function is the total probability of finding the system on the particular surface of constant $A(X)$ labeled by the parameter a. Note that because of the delta function, the integral of g over a is normalized to unity,

$$\int_{\text{all}} da\, g(a, t) = 1. \tag{9.36}$$

The exact time dependence of distribution functions is determined by the Liouville equation, which describes motion of a system in phase space X. Reduced distribution functions contain only the parameters a and do not refer directly to phase space. Changing to a reduced

NONLINEAR PROBLEMS 177

description of the dynamics requires careful attention to the distinction between functions of X and variables that do not depend on X. When dealing with a set of dynamical variables, we denote the set itself by the uppercase vector $\mathbf{A}$ and its numerical values by the lowercase $\mathbf{a}$. Often, for brevity, we omit the phase-space position X from $\mathbf{A}(X)$. The delta function $\delta(\mathbf{A} - \mathbf{a})$ denotes the product of delta functions for each individual element of the set.

There are two ways to proceed. In one, a projection operator is used to separate the Liouville equation for the phase space distribution function into relevant and irrelevant parts. An alternative procedure, the one to be followed here, makes use of the generalized Langevin equation derived in section 8.2. The idea is to replace the original variable $\mathbf{A}$ by a new dynamical variable $G(\mathbf{a})$, indexed by the parameters $\mathbf{a}$,

$$G(\mathbf{a}) = \delta(\mathbf{A}(X) - \mathbf{a}). \tag{9.37}$$

Because $\mathbf{A}$ is a function of the phase point X, G is defined in the complete phase space. This new variable carries information about all powers of $\mathbf{A}$. At time t, this variable is $G(\mathbf{a}, t) = \delta(\mathbf{A}(t) - \mathbf{a})$, and its average over a given initial phase space distribution is

$$\langle G(\mathbf{a}, t) \rangle_0 = \int dX \delta(\mathbf{A}(t) - \mathbf{a}) f(X, 0)$$
$$= \int dX \delta(\mathbf{A} - \mathbf{a}) f(X, t) = g(\mathbf{a}, t). \tag{9.38}$$

The desired reduced distribution function is the phase space average of the dynamical variable G. The formal calculation of such quantities is precisely what the generalized Langevin equation is good for.

The Derivation

The only significant modification of the derivation of Langevin equations given earlier is to replace $\mathbf{A}$ by $\delta(\mathbf{A} - \mathbf{a})$. We do not subtract the equilibrium average of the variable, but keep the whole quantity. This average is

$$\langle G(\mathbf{a}) \rangle_{eq} = g_{eq}(\mathbf{a}). \tag{9.39}$$

The inner product (,) of two different variables (taking advantage of properties of delta functions) is

$$(G(\mathbf{a}), G(\mathbf{a}')) = \langle \delta(\mathbf{A} - \mathbf{a}) \delta(\mathbf{A} - \mathbf{a}') \rangle_{eq} = g_{eq}(\mathbf{a}) \delta(\mathbf{a} - \mathbf{a}'). \tag{9.40}$$

This is diagonal in the indices $\mathbf{a}$ and $\mathbf{a}'$, and so its inverse is

$$\left((G,G)^{-1}\right)_{aa'} = \frac{1}{g_{eq}(\mathbf{a})}\delta(\mathbf{a}-\mathbf{a}'). \tag{9.41}$$

The projection operator is defined by

$$PB = \iint d\mathbf{a}\,d\mathbf{a}'(B,G(\mathbf{a}))\left((G,G)^{-1}\right)_{\mathbf{a},\mathbf{a}'}G(\mathbf{a}'). \tag{9.42}$$

For brevity of notation, it is convenient to introduce the conditional equilibrium distribution $f_{eq}(X;\mathbf{a})$ in phase space. This is distinguished from the complete equilibrium distribution by explicitly including the parameters $\mathbf{a}$. The conditional distribution is made by selecting out of the complete equilibrium distribution only those phase points that lie on the specified surface $\mathbf{A}(X) = \mathbf{a}$. It has the general form

$$f_{eq}(X;\mathbf{a}) = \delta(\mathbf{A}-\mathbf{a})\frac{f_{eq}(X)}{g_{eq}(\mathbf{a})} = G(\mathbf{a})\frac{f_{eq}(X)}{g_{eq}(\mathbf{a})}. \tag{9.43}$$

The denominator provides phase space normalization for the conditional distribution,

$$\int dX f_{eq}(X;\mathbf{a}) = 1 \quad \text{(all } \mathbf{a}\text{)}. \tag{9.44}$$

The average of any B with this distribution will be denoted $\langle B;\mathbf{a}\rangle$,

$$\langle B;\mathbf{a}\rangle = \int dX B(X) f_{eq}(X;\mathbf{a}). \tag{9.45}$$

Then the projection of any B is simply

$$PB = \int d\mathbf{a}\langle B;\mathbf{a}\rangle G(\mathbf{a}). \tag{9.46}$$

On using the results of section 8.2, the generalized Langevin equation for $G(\mathbf{a},t)$ is

$$\frac{\partial}{\partial t}G(\mathbf{a},t)$$
$$= \int d\mathbf{a}'\, i\Omega_{aa'}G(\mathbf{a}',t) - \int_0^t ds \int d\mathbf{a}'\, K_{aa'}(s)G(\mathbf{a}',t-s) + F(\mathbf{a},t). \tag{9.47}$$

The streaming term $i\Omega$ has the general form

$$i\Omega_{aa'} = \int d\mathbf{a}''(LG(\mathbf{a}),G(\mathbf{a}''))\left((G,G)^{-1}\right)_{a''a'}. \tag{9.48}$$

With the notation of conditional averages, this is

$$i\Omega_{aa'} = \langle LG(\mathbf{a});\mathbf{a}'\rangle. \tag{9.49}$$

NONLINEAR PROBLEMS 179

The Liouville operator is a first-order differential operator in the space X, and we can use the chain rule of differentiation,

$$LG(\mathbf{a}) = \frac{\partial \delta(\mathbf{A}-\mathbf{a})}{\partial \mathbf{A}} \cdot L\mathbf{A} = -\frac{\partial}{\partial \mathbf{a}} \cdot \delta(\mathbf{A}-\mathbf{a})L\mathbf{A}. \tag{9.50}$$

When the conditional average is put in explicitly, $i\Omega$ is

$$i\Omega_{\mathbf{aa}'} = -\frac{\partial}{\partial \mathbf{a}} \cdot \int dX (L\mathbf{A}) \delta(\mathbf{A}-\mathbf{a}) \delta(\mathbf{A}-\mathbf{a}') f_{eq}(X) \frac{1}{g_{eq}(\mathbf{a}')}. \tag{9.51}$$

Because $\delta(\mathbf{A}-\mathbf{a})\delta(\mathbf{A}-\mathbf{a}') = \delta(\mathbf{A}-\mathbf{a})\delta(\mathbf{a}-\mathbf{a}')$, we can rewrite this as

$$i\Omega_{\mathbf{aa}'} = -\frac{\partial}{\partial \mathbf{a}} \cdot \left(\int dX (L\mathbf{A}) \delta(\mathbf{A}-\mathbf{a}) \frac{f_{eq}}{g_{eq}(\mathbf{a})} \right) \delta(\mathbf{a}-\mathbf{a}'). \tag{9.52}$$

This contains the average of the rate of change $L\mathbf{A}$ in the conditional equilibrium ensemble, which will be denoted by $\mathbf{V}(\mathbf{a})$,

$$\mathbf{V}(\mathbf{a}) = \int dX (L\mathbf{A}) f_{eq}(X;\mathbf{a}) = \langle L\mathbf{A}; \mathbf{a} \rangle. \tag{9.53}$$

So the first term in the right hand side of the Fokker-Planck equation has the standard form,

$$\int d\mathbf{a}' i\Omega_{\mathbf{aa}'} G(\mathbf{a}',t) = -\frac{\partial}{\partial \mathbf{a}} \cdot (\mathbf{V}(\mathbf{a}) G(\mathbf{a},t)), \tag{9.54}$$

and eq. (9.53) is the statistical mechanical formula for $\mathbf{V}$. (Note that this is only part of the full $\mathbf{v}(\mathbf{a})$ in eq. (9.29). There are further contributions from the equilibrium distribution g_{eq}.)

Now we turn to the memory kernel and follow the same procedure that we used to get from $i\Omega_{\mathbf{a},\mathbf{a}'}$ to $\mathbf{V}(\mathbf{a})$,

$$K_{\mathbf{aa}'}(t) = \int d\mathbf{a}'' (e^{(1-P)Lt}(1-P)LG(\mathbf{a}), LG(\mathbf{a}'')) \left((G,G)^{-1}\right)_{\mathbf{a}''\mathbf{a}'}$$

$$= (e^{(1-P)Lt}(1-P)LG(\mathbf{a}), LG(\mathbf{a}')) \frac{1}{g_{eq}(\mathbf{a}')}$$

$$= \left(e^{(1-P)Lt}(1-P)\frac{\partial}{\partial \mathbf{a}} \cdot L\mathbf{A}\delta(\mathbf{A}-\mathbf{a}), \frac{\partial}{\partial \mathbf{a}'} \cdot L\mathbf{A}\delta(\mathbf{A}-\mathbf{a}') \right) \frac{1}{g_{eq}(\mathbf{a}')}. \tag{9.55}$$

Then after integrating by parts over $\mathbf{a}'$, we get

$$\int d\mathbf{a}' K_{\mathbf{aa}'}(s) G(\mathbf{a}', t-s) = -\frac{\partial}{\partial \mathbf{a}} \cdot \int d\mathbf{a}'$$

$$(e^{(1-P)Ls}(1-P)L\mathbf{A}\delta(\mathbf{A}-\mathbf{a}), L\mathbf{A}\delta(\mathbf{A}-\mathbf{a}')) \cdot \frac{\partial}{\partial \mathbf{a}'} \frac{G(\mathbf{a}', t-s)}{g_{eq}(\mathbf{a}')}. \qquad (9.56)$$

At this point, it is convenient to use the abbreviation

$$\mathbf{B}(\mathbf{a}, \mathbf{a}', s) = (e^{(1-P)Ls}(1-P)L\mathbf{A}\delta(\mathbf{A}-\mathbf{a}), L\mathbf{A}\delta(\mathbf{A}-\mathbf{a}')) \frac{1}{g_{eq}(\mathbf{a}')}. \qquad (9.57)$$

Now eq. (9.47) becomes

$$\frac{\partial}{\partial t} G(\mathbf{a}, t) = -\frac{\partial}{\partial \mathbf{a}} \cdot \mathbf{V}(\mathbf{a}) G(\mathbf{a}, t)$$

$$+ \frac{\partial}{\partial \mathbf{a}} \cdot \int_0^t ds \int d\mathbf{a}' \mathbf{B}(\mathbf{a}, \mathbf{a}', s) \cdot g_{eq}(\mathbf{a}') \frac{\partial}{\partial \mathbf{a}'} \frac{G(\mathbf{a}', t-s)}{g_{eq}(\mathbf{a}')} + F(\mathbf{a}, t). \qquad (9.58)$$

This is clearly a complicated equation and not much use in its present form. We will see that it becomes much simpler if the dynamical variables **A** are "slow" in the sense that was discussed earlier.

Before going to the slow variable limit, however, we can derive from this equation both a nonlinear Langevin equation and its corresponding Fokker-Planck equation. To get the Langevin equation, we just multiply eq. (9.58) by **a** and integrate over all **a**, using

$$\mathbf{A}(t) = \int d\mathbf{a}\, \mathbf{a}\, G(\mathbf{A}(t) - \mathbf{a}). \qquad (9.59)$$

Several integrations by parts lead to

$$\frac{\partial}{\partial t} \mathbf{A}(t) = \int d\mathbf{a}\, \mathbf{V}(\mathbf{a}) G(\mathbf{a}, t)$$

$$+ \int d\mathbf{a} \int d\mathbf{a}' \int_0^t ds \left[\frac{1}{g_{eq}(\mathbf{a}')} \frac{\partial}{\partial \mathbf{a}'} g_{eq}(\mathbf{a}') \mathbf{B}(\mathbf{a}, \mathbf{a}', s) \right] G(\mathbf{a}', t-s)$$

$$+ \int d\mathbf{a}\, \mathbf{a}\, F(\mathbf{a}, t). \qquad (9.60)$$

Because G is a delta function, the first term on the right hand side is just $\mathbf{V}(\mathbf{A}(t))$. In the second term, because G is a delta function of the difference between $\mathbf{A}(t-s)$ and $\mathbf{a}'$, integration over $\mathbf{a}'$ produces a function of $\mathbf{A}(t-s)$. The third term is a function of t. Without writing this out in greater detail, this is clearly a generalized nonlinear Langevin equation.

But we can also use eq. (9.58) to get the corresponding Fokker-Planck equation. This merely involves averaging over an initial phase space distribution. The average $\langle G(\mathbf{a}, t) \rangle$ becomes $g(\mathbf{a}, t)$. If the initial distribution is one of constrained equilibrium,

$$f(X, 0) = \int d\mathbf{a}\, f_{eq}(X, \mathbf{a}) g(\mathbf{a}, 0), \tag{9.61}$$

then $(1 - P)f(X, 0) = 0$, and the average of the noise term $\mathbf{F}(\mathbf{a}, t)$ vanishes. The result is a generalized Fokker-Planck equation (R. Zwanzig, 1961).

9.3 Nonlinear Langevin Equations and Fokker-Planck Equations for Slow Variables

As in the discussion of linear Langevin equations, the nonlinear Langevin equation and Fokker-Planck equation derived in the last section are considerably simpler if the variables $\mathbf{A}$ are "slow." Then $L\mathbf{A}$ contains a smallness parameter λ, so that $L\mathbf{A}$ is of the order of λ. The streaming velocity $\mathbf{v}$, containing one factor of $L\mathbf{A}$, is of order λ, and the diffusion coefficient $\mathbf{B}$, containing two factors of $L\mathbf{A}$, is formally of order λ^2. We have seen that to order λ^2, the operator $\exp[(1 - P)Ls]$ can be replaced by $\exp[Ls]$, leading to a more conventional time correlation function,

$$\mathbf{B}(\mathbf{a}, \mathbf{a}', s) = (e^{Ls}(1 - P)L\mathbf{A}\delta(\mathbf{A} - \mathbf{a}), L\mathbf{A}\delta(\mathbf{A} - \mathbf{a}'))\frac{1}{g_{eq}(\mathbf{a}')} + O(\lambda^3). \tag{9.62}$$

The quantity $(1 - P)L\mathbf{A}\,\delta(\mathbf{A} - \mathbf{a})$ has the very simple form,

$$(1 - P)L\mathbf{A}\delta(\mathbf{A} - \mathbf{a}) = (L\mathbf{A} - \mathbf{V}(\mathbf{a}))\delta(\mathbf{A} - \mathbf{a}); \tag{9.63}$$

it contains the fluctuation of the actual rate of change $L\mathbf{A}$ from its conditional equilibrium average $\mathbf{V}$.

The time scale for appreciable changes in the dynamical variables $\mathbf{A}(X, t)$ is of the order of $1/\lambda$ and can be very long when λ is small. The decay time of the time correlation function is determined by the exact Liouville operator and has nothing to do with λ. The slow change in $\mathbf{A}$ allows us to replace $\mathbf{A}(X, s)$ by $\mathbf{A}(X, 0) = \mathbf{A}$ in the delta function,

$$\mathbf{B}(\mathbf{a}, \mathbf{a}', s) = ([L\mathbf{A}(s) - \mathbf{V}(\mathbf{a})]\delta(\mathbf{A} - \mathbf{a}), L\mathbf{A}\delta(\mathbf{A} - \mathbf{a}'))$$
$$\frac{1}{g_{eq}(\mathbf{a}')} + O(\lambda^3). \tag{9.64}$$

The inner product contains a product of two delta functions; then we can replace one of them, $\delta(\mathbf{A} - \mathbf{a}')$, by $\delta(\mathbf{a} - \mathbf{a}')$, and rewrite the inner product as a conditional equilibrium average,

$$\mathbf{B}(\mathbf{a}, \mathbf{a}', s) = \langle [L\mathbf{A}(s) - \mathbf{V}(\mathbf{a})]L\mathbf{A}; \mathbf{a}\rangle \delta(\mathbf{a} - \mathbf{a}') + O(\lambda^3). \tag{9.65}$$

The diffusion coefficient for slow variables is local in $\mathbf{a}$-space. The coefficient of the delta function will be denoted by $\mathbf{B}(\mathbf{a}, s)$ with a single $\mathbf{a}$, and the term of order λ^3 will be omitted. This quantity is

$$\mathbf{B}(\mathbf{a}, s) = \langle [L\mathbf{A}(s) - \mathbf{V}(\mathbf{a})]L\mathbf{A}; \mathbf{a}\rangle. \tag{9.66}$$

The second factor $L\mathbf{A}$ can be replaced by $L\mathbf{A} - \mathbf{V}(\mathbf{a})$ without any effect on $\mathbf{B}$.

In the slow variable limit, eq. (9.58) becomes

$$\frac{\partial}{\partial t} G(\mathbf{a}, t) = -\frac{\partial}{\partial \mathbf{a}} \cdot (\mathbf{V}(\mathbf{a})G(\mathbf{a}, t))$$
$$+ \int_0^t ds \frac{\partial}{\partial \mathbf{a}} \cdot \mathbf{B}(\mathbf{a}, s) \cdot g_{eq}(\mathbf{a}) \frac{\partial}{\partial \mathbf{a}} \frac{G(\mathbf{a}, t-s)}{g_{eq}(\mathbf{a})} + F(\mathbf{a}, t). \tag{9.67}$$

Finally, we note that if we are not interested in fine details of time dependence, the slow variable limit allows us to make a Markovian approximation; the time-dependent kernel decays rapidly on the time scale of the slow variables. Then $G(t - s)$ at time $t - s$ is essentially the same as $G(t)$ at time t, and we need only the time integral of the kernel. This will be denoted by $\mathbf{B}(\mathbf{a})$ without the variable s,

$$\mathbf{B}(\mathbf{a}) = \int_0^\infty ds \mathbf{B}(\mathbf{a}, s). \tag{9.68}$$

By making a Markovian approximation, we obtain

$$\frac{\partial}{\partial t} G = -\frac{\partial}{\partial \mathbf{a}} \cdot (\mathbf{V}(\mathbf{a})G) + \frac{\partial}{\partial \mathbf{a}} \cdot \mathbf{B}(\mathbf{a}) \cdot g_{eq} \frac{\partial}{\partial \mathbf{a}} \frac{G}{g_{eq}} + F(\mathbf{a}, t) + O(\lambda^3), \tag{9.69}$$

with statistical mechanical expressions for the coefficients,

$$\mathbf{V}(\mathbf{a}) = \langle L\mathbf{A}; \mathbf{a}\rangle, \tag{9.70}$$

$$\mathbf{B}(\mathbf{a}) = \int_0^\infty ds \langle (L\mathbf{A}(s) - \mathbf{V}(\mathbf{a}))(L\mathbf{A} - \mathbf{V}(\mathbf{a})); \mathbf{a}\rangle. \tag{9.71}$$

$\mathbf{B}(\mathbf{a})$ is the integral of the time correlation function of the fluctuation of the rate of change of $\mathbf{A}$, $L\mathbf{A} - \mathbf{V}(\mathbf{a})$, calculated in the conditional equilibrium ensemble. Both $\mathbf{V}$ and $\mathbf{B}$ can depend on the variables $\mathbf{a}$.

NONLINEAR PROBLEMS 183

The result of these manipulations, eq. (9.69), is an approximate Langevin equation for the dynamical variable $G = \delta(\mathbf{A} - \mathbf{a})$. The *average* of this equation over some constrained equilibrium initial distribution is a Fokker-Planck equation,

$$\frac{\partial}{\partial t} g = -\frac{\partial}{\partial \mathbf{a}} \cdot (\mathbf{V}(\mathbf{a})g) + \frac{\partial}{\partial \mathbf{a}} \cdot \mathbf{B}(\mathbf{a}) \cdot g_{eq} \frac{\partial}{\partial \mathbf{a}} \frac{g}{g_{eq}}. \qquad (9.72)$$

This is the main result of this section.

The Langevin equation for the complete G can be used to derive a nonlinear Langevin equation for the first power $\mathbf{A}$ by means of

$$A = \int d\mathbf{a}\, \mathbf{a} G(\mathbf{a}) = \int d\mathbf{a}\, \mathbf{a} \delta(\mathbf{A} - \mathbf{a}). \qquad (9.73)$$

The result is

$$\frac{\partial}{\partial t} \mathbf{A} = \mathbf{V}(\mathbf{A}) + \frac{\partial}{\partial \mathbf{A}} \cdot \mathbf{B}(\mathbf{A}) + \mathbf{B}(\mathbf{A}) \cdot \frac{\partial}{\partial \mathbf{A}} \ln g_{eq}(\mathbf{A}) + \mathbf{F}(t). \qquad (9.74)$$

So the streaming velocity in eq. (9.62) has three parts,

$$\mathbf{v}(\mathbf{a}) = \mathbf{V}(\mathbf{a}) + \frac{\partial}{\partial \mathbf{a}} \cdot \mathbf{B}(\mathbf{a}) + \mathbf{B}(\mathbf{a}) \cdot \frac{\partial}{\partial \mathbf{a}} \ln g_{eq}(\mathbf{a}), \qquad (9.75)$$

where $\mathbf{a}$ is to be replaced by $\mathbf{A}(X)$. Note that the quantity

$$\mathbf{F}_{thermo}(\mathbf{a}) = \frac{\partial}{\partial \mathbf{a}} \ln g_{eq}(\mathbf{a}) \qquad (9.76)$$

can be regarded as a thermodynamic force driving the system to equilibrium. The noise is

$$\mathbf{F}(t) = \int d\mathbf{a}\, \mathbf{a} \mathbf{F}(\mathbf{a}, t). \qquad (9.77)$$

Illustration

Calculating the functions $\mathbf{V}$ and $\mathbf{B}$ is not generally easy. An exception is for the special model of Brownian motion discussed earlier, a nonlinear system interacting with a harmonic oscillator heat bath. The same notation will be used here. The dynamical variable $\mathbf{A}$ and its rate of change $L\mathbf{A}$ are the two-component vectors

$$\mathbf{A} = \begin{pmatrix} x \\ p \end{pmatrix}, \quad L\mathbf{A} = \begin{pmatrix} p/m \\ F(x) + \sum_j \gamma_j \left(q_j - \frac{\gamma_j}{\omega_j^2} x \right) \end{pmatrix}. \qquad (9.78)$$

The average of L**A** in the constrained equilibrium ensemble, that is, an average over the heat bath variables q_j and p_j for fixed x and p, is

$$\mathbf{V}(\mathbf{a}) = \langle L\mathbf{A}, \mathbf{a} \rangle = \begin{pmatrix} p/m \\ F(x) \end{pmatrix}. \tag{9.79}$$

So the streaming term in the Fokker-Planck equation is

$$\frac{\partial}{\partial \mathbf{a}} \cdot \mathbf{V}(\mathbf{a}) g(\mathbf{a}, t) = \left(\frac{\partial}{\partial x}, \frac{\partial}{\partial p} \right) \cdot \begin{pmatrix} p/m \\ F(x) \end{pmatrix} g = \frac{p}{m} \frac{\partial g}{\partial x} + F(x) \frac{\partial g}{\partial p}. \tag{9.80}$$

The fluctuating part of L**A** is the vector

$$L\mathbf{A} - \mathbf{V}(\mathbf{a}) = \left(\sum \gamma_j \left(q_j - \frac{\gamma_j}{\omega_j^2} x \right) \right). \tag{9.81}$$

The time correlation function of the fluctuation is the square matrix

$$\mathbf{B}(\mathbf{a}, s) = \begin{pmatrix} 0 & 0 \\ 0 & B_{pp} \end{pmatrix}$$

$$B_{pp} = \left\langle \sum \gamma_j \left(q_j(s) - \frac{\gamma_j}{\omega_j^2} x \right) \sum \gamma_k \left(q_k - \frac{\gamma_k}{\omega_k^2} x \right); (x, p) \right\rangle. \tag{9.82}$$

Again, the average is taken over the heat bath variables at fixed x and p and as earlier the result is

$$B_{pp}(s) = \sum \frac{\gamma_j^2}{\omega_j^2} \cos(\omega_j s) = K(s). \tag{9.83}$$

The (p, p) element of **B** is precisely the memory function $K(s)$ that was derived there. The diffusion coefficient B is the time integral of K, and the Fokker-Planck equation (the x-dependent part of g_{eq} cancels out) is

$$\frac{\partial g}{\partial t} = -\frac{p}{m}\frac{\partial g}{\partial x} - F(x)\frac{\partial g}{\partial p} + \frac{\partial}{\partial p} B e^{-\beta p^2/2m} \frac{\partial}{\partial p} e^{\beta p^2/2m} g$$

$$= -\frac{p}{m}\frac{\partial g}{\partial x} - F(x)\frac{\partial g}{\partial p} + \frac{\partial}{\partial p} B \left(\frac{\beta}{m} pg + \frac{\partial}{\partial p} g \right). \tag{9.84}$$

Noise and Initial States

The nonlinear Langevin equation derived in this section contains a noise that is orthogonal to all functions of the variables A, so that it averages to zero if the initial distribution is determined entirely by some function of A. In the same way, the associated Fokker-Planck equation obtained by averaging a distribution function over noise is valid only for the same kind of initial distribution. A linear Langevin equation for A was derived in section 8.2,

$$\frac{\partial}{\partial t} A(t) = i\Omega A(t) - \int_0^t ds K_L(s) A(t-s) + F_L(t). \tag{9.85}$$

The subscript $_L$ is used to distinguish the memory function and noise from those appearing in the nonlinear equations. This equation is generally exact, since it is a formal rearrangement of the Liouville equation; but it is useful only if the average of the noise over an initial distribution vanishes. This limits its applicability to initial states that are very close to equilibrium. An illustration of the relationship of linear to nonlinear Langevin equations was given in section 8.5.

9.4 Kinds of Nonlinearity

The Langevin equation for slow variables that was discussed in the preceding chapter contains three functions of the chosen variables, a streaming velocity $\mathbf{V}(\mathbf{a})$, a diffusion coefficient $\mathbf{B}(\mathbf{a})$, and an equilibrium distribution $g_{eq}(\mathbf{a})$. Each of these can contribute nonlinear terms to the Langevin equation.

The equilibrium distribution may have a single minimum as a function of $\mathbf{a}$; in this case, we can usually expand about the minimum, and then the distribution is approximately Gaussian,

$$g_{eq}(\mathbf{a}) \propto \exp\left(-\frac{1}{2}\mathbf{a}\cdot\mathbf{M}^{-1}\cdot\mathbf{a}\right), \quad \mathbf{M} = \langle \mathbf{aa}\rangle_{eq}. \tag{9.86}$$

Or, the distribution may have several minima; then we are likely to be dealing with some reactive barrier crossing problem or with a phase transition. In the following discussion, we consider for simplicity only the Gaussian distribution.

The streaming velocity $\mathbf{V}(\mathbf{a})$ is given by the constrained average

$$V_i(\mathbf{a}) = \langle LA_i; \mathbf{a}\rangle = \frac{\langle LA_i \delta(\mathbf{A}-\mathbf{a})\rangle_{eq}}{\langle \delta(\mathbf{A}-\mathbf{a})\rangle_{eq}}. \tag{9.87}$$

The denominator is the equilibrium distribution g_{eq}. The constrained average is awkward to deal with. However, we can derive its series expansion by taking moments of $\mathbf{V}$. The first three are

$$\int d\mathbf{a} V_i(\mathbf{a}) g_{eq}(\mathbf{a}) = \langle V_i\rangle_{eq} = \langle LA_i\rangle_{eq} = 0 \tag{9.88}$$

$$\int d\mathbf{a} V_i(\mathbf{a}) a_j g_{eq}(\mathbf{a}) = \langle V_i a_j\rangle_{eq} = \langle (LA_i) A_j\rangle_{eq} \tag{9.89}$$

186 NONEQUILIBRIUM STATISTICAL MECHANICS

$$\int d\mathbf{a} \mathbf{V}_i(\mathbf{a}) a_j a_k g_{eq}(\mathbf{a}) = \langle \mathbf{V}_i a_j a_k \rangle_{eq} = \langle (LA_i) A_j A_k \rangle_{eq}. \tag{9.90}$$

The expansion of **V** starts out with linear and quadratic terms,

$$\mathbf{V}_i(\mathbf{a}) = \sum_j V_{ij}^{(1)} a_j + \sum_{j,k} V_{ijk}^{(2)} a_j a_k + \cdots. \tag{9.91}$$

The coefficients are determined by the moments. The zeroth moment leads to a condition that $V_{ijk}^{(2)}$ must satisfy

$$\langle \mathbf{V}_i \rangle_{eq} = \sum_{j,k} V_{ijk}^{(2)} M_{jk} = 0. \tag{9.92}$$

The first moment leads to

$$\langle \mathbf{V}_i a_j \rangle_{eq} = \sum_k V_{ik}^{(1)} M_{kj} = \langle (LA_i) A_j \rangle_{eq}. \tag{9.93}$$

This is the quantity that appears in linear response theory,

$$V^{(1)} = i\Omega = (LA, A) \cdot (A, A)^{-1}. \tag{9.94}$$

The second moment is

$$\langle \mathbf{V}_i a_j a_k \rangle_{eq} = \sum_{l,m} V_{ilm}^{(2)} \langle a_l a_m a_j a_k \rangle_{eq} = \langle (LA_i) A_j A_k \rangle_{eq}. \tag{9.95}$$

We use the Gaussian character of g_{eq} to relate the fourth moment to second moments,

$$\sum_{l,m} V_{ilm}^{(2)} \langle a_l a_m a_j a_k \rangle_{eq} = \sum_{l,m} V_{ilm}^{(2)} (M_{lm} M_{jk} + M_{lj} M_{mk} + M_{lk} M_{mj})$$

$$= 2 \sum_{l,m} V_{ilm}^{(2)} M_{mk} M_{lj}. \tag{9.96}$$

so that on multiplying by inverses of **M** we find

$$V_{ijk}^{(2)} = \frac{1}{2} \langle (LA_i)(\mathbf{A} \cdot \mathbf{M}^{-1})_j (\mathbf{A} \cdot \mathbf{M}^{-1})_k \rangle_{eq}. \tag{9.97}$$

Quantities like these will appear in the treatment of mode-coupling theory to be presented shortly.

The same general procedure can be used to find an expansion of **B(a)** in powers of **a**,

$$B_{ij}(\mathbf{a}) = B_{ij}^{(0)} + \sum_k B_{ijk}^{(1)} a_k + \cdots. \tag{9.98}$$

The zeroth-order term, independent of **a**, is proportional to the rate matrix **K** that appears in linear response theory,

NONLINEAR PROBLEMS

$$\mathbf{B}^{(0)} = \int_0^\infty dt \langle (L\mathbf{A}(t) - i\Omega \cdot \mathbf{a})(L\mathbf{A} - i\Omega \cdot \mathbf{a}) \rangle_{eq} = \mathbf{K} \cdot \mathbf{M}. \tag{9.99}$$

Very little is known about higher order terms in this expansion. In the following, we will assume that $\mathbf{B}$ is approximated by $\mathbf{B}^{(0)}$, and we omit the superscript.

When the equilibrium distribution is Gaussian, and $\mathbf{F}$ is Gaussian white noise, the Langevin equation is

$$\frac{\partial}{\partial t}\mathbf{a} = \mathbf{V}(\mathbf{a}) - \mathbf{K} \cdot \mathbf{a} + \mathbf{F}(t). \tag{9.100}$$

The corresponding Fokker-Planck equation is

$$\frac{\partial}{\partial t} g = -\frac{\partial}{\partial \mathbf{a}} \cdot (\mathbf{V}g) + \frac{\partial}{\partial \mathbf{a}} \cdot (\mathbf{K} \cdot \mathbf{a}g) + \frac{\partial}{\partial \mathbf{a}} \cdot \mathbf{B} \cdot \frac{\partial}{\partial \mathbf{a}} g. \tag{9.101}$$

In many applications, the nonlinear part of $\mathbf{V}$ comes from a convective term, for example, $\mathbf{v} \cdot \nabla \mathbf{v}$ in the Navier-Stokes equation or $\nabla \cdot \mathbf{v}C$ in the diffusion equation. From here on, we consider only nonlinearities that are quadratic in $\mathbf{a}$, as in eq. (9.91). As in earlier discussions of Fokker-Planck equations, the linear part of $\mathbf{V}$ is combined with the dissipative term,

$$\Theta = i\Omega - \mathbf{K}. \tag{9.102}$$

The nonlinear part is denoted by $\delta \mathbf{V}$,

$$\delta \mathbf{V} = \mathbf{V} - i\Omega \cdot \mathbf{a}. \tag{9.103}$$

The Fokker-Planck equation may be separated into linear and nonlinear parts,

$$\frac{\partial}{\partial t} g = (\mathcal{D}_0 + \mathcal{D}_1)g, \tag{9.104}$$

where the linear part is

$$\mathcal{D}_0 g = -\frac{\partial}{\partial \mathbf{a}} \cdot (\Theta \cdot \mathbf{a}g) + \frac{\partial}{\partial \mathbf{a}} \cdot \mathbf{B} \cdot \frac{\partial}{\partial \mathbf{a}} g, \tag{9.105}$$

and the nonlinear part is

$$\mathcal{D}_1 g = -\frac{\partial}{\partial \mathbf{a}} \cdot (\delta \mathbf{V} g). \tag{9.106}$$

The Green's function for the linear part,

$$\frac{\partial}{\partial t} G(\mathbf{a}, t | \mathbf{a}_0) = \mathcal{D}_0 G(\mathbf{a}, t | \mathbf{a}_0), \qquad G(\mathbf{a}, 0 | \mathbf{a}_0) = \delta(\mathbf{a} - \mathbf{a}_0), \tag{9.107}$$

was derived in section 2.3.

9.5 Nonlinear Transport Equations

Here we show how nonlinear Langevin equations can lead to nonlinear transport equations for the averages of dynamical variables and to a "fluctuation-renormalization" of transport coefficients.

As before, the dynamical variables are denoted by $\mathbf{a} = \{a_1, a_2, \ldots\}$. To keep the algebra as simple as possible, we assume that the dynamical variables have already been orthogonalized and normalized to unity,

$$M_{jk} = \langle a_j a_k \rangle = \delta_{jk}. \tag{9.108}$$

The equilibrium distribution is Gaussian. We assume that the streaming velocity,

$$\mathbf{V}(\mathbf{a}) = \langle L\mathbf{A}; \mathbf{a} \rangle, \tag{9.109}$$

has the familiar linear part and a quadratic nonlinearity $\delta \mathbf{V}$,

$$\delta V_i(\mathbf{a}) = \sum_{jk} V_{ijk} a_j a_k. \tag{9.110}$$

The coefficients V_{ijk} are given by eq. (9.97),

$$V_{ijk} = \frac{1}{2} \langle (LA_i) A_j A_k \rangle_{eq}. \tag{9.111}$$

The coefficients satisfy the identity

$$V_{ijk} + V_{jki} + V_{kij} = \frac{1}{2} \langle L(A_i A_j A_k) \rangle_{eq} = 0. \tag{9.112}$$

The Langevin equation is

$$\frac{\partial}{\partial t} a_j = \sum_k \Theta_{jk} a_k + \sum_{kl} V_{jkl} a_k a_l + F_j(t), \tag{9.113}$$

and the corresponding Fokker-Planck equation is

$$\frac{\partial}{\partial t}g = -\sum_{jk}\frac{\partial}{\partial a_j}(\Theta_{jk}a_k g) - \sum_{jkl}\frac{\partial}{\partial a_j}(V_{jkl}a_k a_l g) + \sum_{jk}\frac{\partial}{\partial a_j}K_{jk}\frac{\partial g}{\partial a_k}.$$
(9.114)

Now we use the Fokker-Planck equation to find an equation for the average $\langle a_j \rangle$,

$$\langle a_j(t) \rangle = \int d\mathbf{a}\, a_j g(\mathbf{a}, t).$$
(9.115)

Then its rate of change is

$$\frac{\partial}{\partial t}\langle a_j(t)\rangle = \sum_k \Theta_{jk}\langle a_k(t)\rangle + \sum_{jk} V_{jkl}\langle a_j a_l \rangle.$$
(9.116)

This contains the average of a product of two *a*s. The error made in replacing the average of a product by a product of the averages depends on the degree of sharpness of the **a**-space distribution $g(\mathbf{a}, t)$. This suggests consideration of the cumulants of the distribution.

Cumulants are defined as follows. Given a distribution, we evaluate its generating function,

$$\Gamma(\xi) = \int d\mathbf{a}\, e^{\xi \cdot \mathbf{a}} g(\mathbf{a}) = \langle e^{\xi \cdot \mathbf{a}} \rangle,$$
(9.117)

which depends on the variables ξ. We take the logarithm of this quantity, denoted by $C(\xi)$, and expand it in powers of the ξs,

$$\ln \Gamma(\xi) = C(\xi) = \sum_j c_j \xi_j + \frac{1}{2}\sum_{jk} c_{jk}\xi_j \xi_k + \frac{1}{3!}\sum_{jkl} c_{jkl}\xi_j \xi_k \xi_l + \cdots.$$
(9.118)

The coefficients $c_j, c_{jk}, c_{jkl}, \ldots$ are called the first, second, third, $\ldots$ cumulants of the distribution. The cumulants are a measure of the shape of the distribution. Since $\Gamma(\xi)$ can be expanded directly in powers of the ξs, with coefficients that are various moments of the distribution, the cumulants are clearly related to the moments; the first three cumulants are

$$c_j = \langle a_j \rangle$$
(9.119)

$$c_{jk} = \langle a_j a_k \rangle - \langle a_j \rangle\langle a_k \rangle$$
(9.110)

$$c_{jkl} = \langle a_j a_k a_l \rangle - \langle a_j a_k \rangle\langle a_l \rangle - \langle a_k a_l \rangle\langle a_j \rangle - \langle a_l a_j \rangle\langle a_k \rangle + 2\langle a_j \rangle\langle a_k \rangle\langle a_l \rangle.$$
(9.121)

Note that if the underlying distribution is Gaussian, the third and higher cumulants all vanish.

Now we return to the solution of the Fokker-Planck equation. By integration we find an equation of motion for the generating function,

$$\frac{\partial \Gamma}{\partial t} = \sum_{jk} \xi_j \Theta_{jk} \frac{\partial \Gamma}{\partial \xi_k} + \sum_{jk} \xi_j K_{jk} \xi_k \Gamma + \sum_{jkl} \xi_j V_{jkl} \frac{\partial^2 \Gamma}{\partial \xi_k \partial \xi_l}. \tag{9.122}$$

This provides an equation for the cumulant generating function,

$$\frac{\partial C}{\partial t} = \sum_{jk} \xi_j \Theta_{jk} \frac{\partial C}{\partial \xi_k} + \sum_{jk} \xi_j K_{jk} \xi_k + \sum_{jkl} \xi_j V_{jkl} \frac{\partial^2 C}{\partial \xi_k \partial \xi_l} + \sum_{jkl} \xi_j V_{jkl} \frac{\partial C}{\partial \xi_k} \frac{\partial C}{\partial \xi_l}. \tag{9.123}$$

Note that a term quadratic in C has appeared. When the function C is expanded in powers of ξ, and terms of the same order are collected, we obtain a series of equations for the various cumulants,

$$\frac{\partial}{\partial t} c_j = \sum_k \Theta_{jk} c_k + \sum_{kl} V_{jkl} c_k c_l + \sum_{kl} V_{jkl} c_{kl} \tag{9.124}$$

$$\frac{\partial}{\partial t} c_{jk} = \sum_l \Theta_{jl} c_{lk} + K_{jk} + \sum_{mn} V_{jmn}(c_{kmn} + c_m c_{nk} + c_n c_{mk}) + \text{transpose}, \tag{9.125}$$

$$\frac{\partial}{\partial t} c_{jkl} = \sum_m [\Theta_{jm} c_{mkl} + \Theta_{km} c_{jml} + \Theta_{lm} c_{jkm}] + O(\delta V). \tag{9.126}$$

The equation for the second cumulant must be symmetrized in j and k, as indicated by "+ transpose." In the equation for the third cumulant, terms containing the nonlinear coupling V_{jkl} have not been written down explicitly; these terms contain the fourth cumulant.

Perturbation Expansion

The goal of the perturbation expansion is to find an equation of motion for the average of the first cumulant or mean value to second order in δV. This means that we have to find the second cumulant to first order in δV. To do this, we first have to estimate the third cumulant. To zeroth order, the differential equations for c_{jkl} are linear equations with constant coefficients. The time dependence of $c_{jkl}(t)$ is determined by the initial value $c_{pqr}(0)$ and by some complicated time-dependent coefficients $I_{jkl,pqr}(t)$ that we do not need to know here. The formal solution is

$$c_{jkl}(t) = \sum_{pqr} I_{jkl,pqr}(t) c_{pqr}(0) + O(\delta V). \tag{9.127}$$

To get the second cumulant, we take advantage of the symmetry $\Omega_{jk} = -\Omega_{kj}$ to rewrite $K_{jk} + K_{kj}$ as

$$K_{jk} + K_{kj} = -\sum_{l}(\Theta_{jl}\delta_{lk} + \Theta_{kl}\delta_{lj}) \tag{9.128}$$

and rewrite eq. (9.126) as

$$\frac{\partial}{\partial t}c_{jk} = \sum_{l}\Theta_{jl}(c_{lk} - \delta_{lk}) + \text{transpose} + O(\delta V). \tag{9.129}$$

This has the matrix operator solution (found by integrating and using the transpose of the matrix Θ on the right),

$$c_{jk}(t) = \delta_{jk} + \int_0^t ds \sum_{mn}(e^{s\Theta})_{jm}(c_{mn}(0) - \delta_{mn})(e^{s\Theta^\dagger})_{nk} + O(\delta V). \tag{9.130}$$

This contains the initial value $c_{mn}(0) - \delta_{mn}$. If the *initial* distribution $g(\mathbf{a}, 0)$ is Gaussian, with arbitrary first cumulant but with equilibrium second cumulant, then the initial value terms in eqs. (9.127) and (9.130) vanish, so that $c_{jk}(t) = \delta_{jk} + O(\delta V)$ and $c_{jkl}(t) = O(\delta V)$. For simplicity, we will emphasize this special case, but we will write *IVT* for "initial value terms" in the various equations that follow, in order to be reminded that such terms can appear.

Next, we substitute the zeroth order values of c_{jk} and c_{jkl} in the $O(\delta V)$ terms of eq. (9.125), and we use the symmetries $V_{jkl} = V_{jlk}$ and $V_{jkl} + V_{ljk} = -V_{ljk}$ to simplify terms. This leads to

$$\frac{\partial}{\partial t}c_{jk} = \sum_{l}\Theta_{jl}(c_{lk} - \delta_{lk}) + \sum_{l}(c_{jl} - \delta_{jl})\Theta_{kl} - 2\sum_{l}c_{l}V_{ljk} + O(\delta V^2) + IVT. \tag{9.131}$$

Now we can find the second cumulant to first order in δV (as before, by integrating and using the transpose of the matrix Θ),

$$c_{jk}(t) = \delta_{jk} - 2\int_0^t ds \sum_{lmn}(e^{(t-s)\Theta})_{jl}V_{nlm}(e^{(t-s)\Theta^\dagger})_{mk}c_n(s) + O(\delta V^2) + IVT. \tag{9.132}$$

If the system is linear, the second cumulant maintains its equilibrium value. To first order in nonlinearity, the second cumulant is driven away from its equilibrium value by the first cumulant. To put this differently, in a nonequilibrium and nonlinear system, correlations that are not present initially must develop in time. Now we substitute this second cumulant in eq. (9.124), which becomes, after some rearrangement,

$$\frac{\partial}{\partial t}c_j(t) = \sum_{k}\Theta_{jk}c_k(t) + \sum_{kl}V_{jkl}c_k(t)c_l(t) - \int_0^t ds \sum_{k}\phi_{jk}(s)c_k(t-s) + IVT, \tag{9.133}$$

192 NONEQUILIBRIUM STATISTICAL MECHANICS

where ϕ_{jm} is an extra memory function,

$$\phi_{jk}(s) = 2 \sum_{lmpq} V_{jlm} V_{kpq} (e^{s\Theta})_{lp} (e^{s\Theta^\dagger})_{qm}. \tag{9.134}$$

The original matrix Θ_{jk} that appears in the linear term has been converted into a new one containing the additional non-Markovian memory function $\phi_{jm}(s)$. This is sometimes called a "fluctuation renormalization" of the "bare" transport coefficient,

$$K_{jk} \to K_{jk} + \int_0^\infty ds\, \phi_{jk}(s). \tag{9.135}$$

If one wants to go beyond this simple second-order perturbation theory, calculations become very much more difficult.

The renormalized transport coefficient derived here is exactly what one gets from the simplest application of the Kadanoff-Swift scheme discussed earlier. The mode-coupling formula for the extra rate constant is

$$\phi_{jk}(t) = \sum_{p \leq q} \sum_{r \leq s} (LA_j, \varphi_{pq})(e^{tL}\varphi_{pq}, \varphi_{rs})(\varphi_{rs}, LA_k). \tag{9.136}$$

The product modes (orthogonal and normalized) are defined by

$$\varphi_{jj} = \frac{1}{\sqrt{2}}(A_j^2 - 1), \qquad \varphi_{jk} = A_j A_k \ (j \neq k). \tag{9.137}$$

As shown in section 9.4, $(LA_i, A_j A_k) = 2V_{ijk}$. After some algebraic manipulation, eq. (9.136) can be converted to eq. (9.134).

10

The Paradoxes of Irreversibility

For many years, certain "paradoxes," usually called the reversibility paradox and the recurrence paradox, have plagued the development of nonequilibrium statistical mechanics. For some time, they led to the belief that there is a fundamental contradiction between Hamiltonian dynamics and the irreversibility that we see everywhere. Even now they are often taken seriously. The subject of this chapter is why there is no fundamental contradiction.

What is meant by irreversibility? Consider a simple experiment. When we put an ice cube into a glass of water, the ice cube melts and the water gets slightly cooler. This experiment has been done countless times, and there have been no reports of the spontaneous reappearance of the ice cube. In human experience, the melting is irreversible.

To be more quantitative, we can use thermodynamic measurements to find the total entropy of both the ice cube and the glass of water before they were brought together, and we can find the total entropy of the final glass of cooler water. We observe that the total entropy has increased. This experiment, and many like it, are summarized in the second law of thermodynamics. In human experience, entropy increases irreversibly.

The paradoxes of irreversibility originated as objections to Boltzmann's H-theorem about the inevitable increase of entropy. The first objection, usually called the "reversibility paradox," was raised by Lord Kelvin and by Loschmidt. In modern terms, this paradox is that the one-way character of irreversibility (the ice cube always melts and never reappears) appears to violate time-reversal symmetry. The fundamental equations of motion of any conservative system are invariant to the substitution of $-t$ for t, or they are symmetric to time-reversal. Consider the dynamical trajectory of the ice cube experiment. At the moment $t = 0$ that the ice cube is put into the glass of water, the dynamical state of the system is a point $\mathbf{X}_0$ in phase space. As the ice melts, this point moves to $\mathbf{X}_t$ at time t. After awhile, the state looks like a glass of cooler water, and no ice cube is present. Now we somehow reverse all velocities, which has the same effect as reversing time. In the same elapsed time t, the state must return to the original $\mathbf{X}_0$, and the ice cube is back. The paradox is that even though there is an initial state where unmelting must occur, we never see it happen. (An interesting exception is the Hahn spin echo experiment. When magnetic fields are present, invariance to time-reversal requires that when the signs of all velocities are changed, the sign of the magnetic field must also be changed. Thus nuclear spin relaxation can be approximately reversed by changing the sign of a magnetic field.)

The second objection was raised by Zermelo and by Poincaré and is usually called the "recurrence paradox." Typically, the motion of any many-body system is confined to the surface of constant energy in phase space. If ergodic theory applies, the trajectory of the system passes, not precisely, but arbitrarily closely, to any assigned position on that surface. Given enough time, it does so arbitrarily often. So any given state of the system will recur to within any assigned accuracy. Any nonequilibrium state that was passed through once will be visited again, or "recur," if one waits long enough. The ice cube should eventually reappear, but we never see it happen.

Understanding the dynamics of an ice cube in a glass of water is not easy. However, we can learn a great deal from analysis of simple models. The coupled harmonic oscillator model of Brownian motion that was discussed in section 1.7 is especially helpful.

Let us perform a thought-experiment, to observe the decay of the velocity v_0 of a particular particle in a one-dimensional harmonic lattice. The experiment starts with a lattice in thermal equilibrium. This means that all initial coordinates and velocities are drawn from an equilibrium ensemble. At $t = 0$, particle $j = 0$ is struck by a neutron, so its initial velocity becomes enormously larger than the thermal velocities of the other particles. By solving the equations of motion, we find that the velocity $v_0(t)$ at time t has two parts. One is proportional to the initial velocity and also to the velocity correlation function. The other part is a linear combination of time-dependent terms proportional to all other initial conditions. It looks like noise. The VCF for this model was found in section 1.7. When all the masses are equal, the VCF of a chain of N particles (with periodic boundary conditions) is

$$C(t) = \frac{1}{N} \sum_{k=0}^{N-1} \cos(\omega_k t), \quad (10.1)$$

where the frequencies are

$$\omega_k = \sin(\pi k/N); \quad (2\sqrt{K/m} = 1). \quad (10.2)$$

In the limit of infinite N, the sum can be replaced by an integral, and the result is the Bessel function $C(t) = J_0(t)$. In the heavy mass limit, where the observed particle is much heavier than the others, $C(t)$ is approximately a decaying exponential function of the absolute value of t.

We consider first the reversibility paradox. The VCF decays to zero at long times, and it is an even function of t, invariant to the replacement of t by $-t$. This means that irreversible decay of the average velocity is compatible with time-reversal symmetry. Why can't one run the trajectory backwards? As the initially large v_0 decays, it affects all the other variables—sound waves are produced that carry away energy and momentum. If we want to reverse the trajectory, we can't just reverse v_0; we have to reverse all the velocities that were developed as a result of the decay of v_0. While this can be done in a computer simulation, it is very hard to do experimentally, since it requires more information than is usually available.

Now we turn to the recurrence paradox. When the number N of particles is finite, the expression for $C(t)$ given in eq. (10.1) is what mathematicians call an "almost periodic function." It is not truly periodic, as long as the frequencies ω_k are incommensurate; however, $C(t)$ will recur to any assigned value c infinitely often. Figure 10.1 shows $\delta C(t) = C(t) - J_0(t)$ for $N = 101$.

The VCF of this 101-particle lattice is very well approximated by the infinite lattice Bessel function up to $t \approx 180$, when the finite size of the chain begins to be felt.

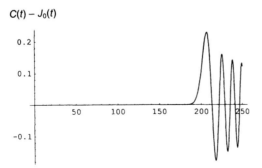

Figure 10.1.1 The deviation of the VCF of a 101 particle chain from the VCF of an infinite chain, as a function of time.

At much longer times, the finite sum appears to behave erratically, as shown in Fig. 10.2, where $500 < t < 1000$.

In this region, the Bessel function is bounded by 0.036. The fluctuations in $C(t)$ are of the order of 0.1.

The mean frequency of return of $C(t)$ to some assigned value c was worked out by M. Kac (1943). Its reciprocal is a mean recurrence time $\tau(c)$. In the equal mass case, the recurrence time is approximately

$$\tau(c) \approx \sqrt{2\pi} e^{Nc^2}. \tag{10.3}$$

In the present illustration, where N is small, typical recurrence times are $\tau(0.1) \approx 12$, and $\tau(0.5) \approx 4 \cdot 10^{11}$. When N is very large, recurrences to the order of $c \approx 1/N^{1/2}$ are frequent. Recurrences to values of the order of N^0 require a time that increases exponentially with N. We may conclude that while recurrences of small fluctuations happen frequently, major recurrences are not likely ever to be seen. What we know about irreversibility is obtained by experiments on a human time scale. The ice cube will eventually reappear, but we won't be around to see it happen.

At least for simple models and for certain initial conditions, irreversible decay is observed over a very long time, and while major recurrences will happen in finite systems, they are highly infrequent. These are natural consequences of equations of motion that have time-reversal symmetry. We don't have to worry about the paradoxes of irreversibility.

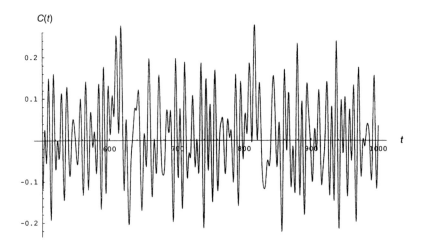

Figure 10.1.2 Apparently random behavior of the VCF of a 101 particle chain at long times, $500 > t > 1000$.

But we still have to worry about the choice of initial conditions. It is possible, in principle, to create initial conditions maliciously, so that bizarre behavior might be observed. However, such conditions would be very hard to accomplish experimentally. In much of the theory discussed in this book, an initial state is drawn from a particular and familiar kind of statistical ensemble. In Brownian motion, the environment is assumed to be in constrained equilibrium. In deriving the quantum mechanical master equation, the initial density matrix is assumed to be diagonal. In Kubo's linear response theory, the initial ensemble is in thermal equilibrium. In deriving generalized Langevin and Fokker-Planck equations, the initial ensemble is assumed to be in constrained equilibrium. In all of these cases, the goal is to separate the time dependence of a dynamical variable into a systematic part and noise. The ensemble average of the noise vanishes, and the ensemble average of the systematic part remains.

So the final issue is to understand why the initial states that we can construct experimentally are characterized by such simple ensembles. We can always assert that an ensemble is just a model of reality that can be confirmed by experiment. This brings us to the essential mystery of statistical mechanics, whether equilibrium or nonequilibrium—why do such models work in the first place?

Appendixes

Appendix 1

First-Order Linear Differential Equations

Many times in this book it is necessary to solve linear inhomogeneous first-order differential equations. The first example appears in section 1.1. The basic rules are reviewed here. The most general equation to be discussed is

$$\frac{dx(t)}{dt} = ax(t) + b(t). \tag{A1.1}$$

There are many ways to derive the solution of this equation. One simple way starts with the substitution

$$x(t) = e^{at} y(t). \tag{A1.2}$$

Then, on taking the time derivative, we get

$$e^{at} \frac{dy(t)}{dt} = b(t) \tag{A1.3}$$

or on multiplying through with $\exp(-at)$,

$$\frac{dy(t)}{dt} = e^{-at} b(t) \tag{A1.4}$$

APPENDIXES 199

Now this is integrated over t. The initial value of y is $y(0) = x(0)$. The result is

$$y(t) = x(0) + \int_0^t ds\, e^{-as} b(s). \tag{A1.5}$$

On using the relation between $y(t)$ and $x(t)$, we obtain the general solution

$$x(t) = e^{at} x(0) + \int_0^t ds\, e^{a(t-s)} b(s). \tag{A1.6}$$

To check this solution, note that $x(t)$ has the right initial value and satisfies the starting differential equation. An alternate form is obtained by replacing s by $t - s$:

$$x(t) = e^{at} x(0) + \int_0^t ds\, e^{as} b(t-s). \tag{A1.7}$$

It is important to note that the symbols in this equation can have various interpretations, and much advantage has been taken of this. The solution was found for the simplest case, where $x(t)$ is a scalar function of t, a is a constant, and $b(t)$ is a given scalar function of t. Suppose that we have to solve a second-order differential equation,

$$\frac{d^2 x}{dt^2} = a_1 x + a_2 \frac{dx}{dt} + b(t). \tag{A1.8}$$

Define the vector function $X(t)$, the matrix A, and the vector B,

$$X(t) = \begin{pmatrix} x(t) \\ \dot{x}(t) \end{pmatrix}, \quad A = \begin{pmatrix} 0 & 1 \\ a_1 & a_2 \end{pmatrix}, \quad B = \begin{pmatrix} 0 \\ b(t) \end{pmatrix}. \tag{A1.9}$$

Then the second-order differential equation becomes a first-order matrix equation,

$$\frac{dX}{dt} = A \cdot X + B(t). \tag{A1.10}$$

This has a solution in the form of eq. (A1.6), but with matrix exponentials,

$$X(t) = e^{At} \cdot X(0) + \int_0^t ds\, e^{A(t-s)} \cdot B(s). \tag{A1.11}$$

An nth-order differential equation with constant coefficients is equivalent to a system of n coupled first-order equations. Then $X(t)$ is a vector function of t, of dimension n; A is a constant matrix of dimension $n \times n$, and $B(t)$ is a vector function of t. The above solution is still formally correct. The derivation in this case follows exactly the same steps, except with vectors and matrices; and they are all kept in their proper places. The symbol exp(tA) is itself a matrix, defined by the series expansion of the exponential.

A third interpretation is that $X(t)$ is an infinite dimensional vector function of t or a function of some auxiliary variables denoted by u; A is a constant infinite dimensional matrix or an operator in the space of u, and $B(t)$ is a function of u and t. The differential equation is now an operator equation. As in the finite matrix case, the above solution is still formally exact, and all operators are in their proper places. Examples arising in this book are where X is either some distribution function or some dynamical observable, and A is a Liouville operator, Fokker-Planck operator, or diffusion operator.

Appendix 2

Gaussian Random Variables

Any quantity A is called a Gaussian random variable if its probability distribution $\rho(a)$, defined by

$$\text{prob}(a < A < a + da) = \rho(a)da, \tag{A2.1}$$

has the Gaussian or normal form,

$$\rho(a) = \frac{1}{\sqrt{2\pi M}} \exp\left(-\frac{1}{2M}(a-\bar{a})^2\right). \tag{A2.2}$$

This distribution has several familiar properties: It is normalized to unity (all integrals are from $-\infty$ to $+\infty$),

$$\int da \rho(a) = 1, \tag{A2.3}$$

the mean value of a is

$$\bar{a} = \int da\, a \rho(a), \tag{A2.4}$$

and the mean squared fluctuation of a is M,

$$M = \int da\, (a-\bar{a})^2 \rho(a). \tag{A2.5}$$

A quantity of special importance is the moment generating function $G(\xi)$,

$$G(\xi) = \int da\, \exp(i\xi a)\rho(a) = \exp\left(i\xi\bar{a} - \frac{1}{2}M\xi^2\right). \tag{A2.6}$$

According to the last equation, the Gaussian distribution function is the inverse Fourier transform of G,

$$\rho(a) = \frac{1}{2\pi} \int d\xi \exp(-i\xi a) G(\xi). \quad (A2.7)$$

All of these familiar statements have a simple extension to a set of Gaussian random variables $\mathbf{A} = \{A_1, A_2, \ldots, A_n\}$. The distribution function $\rho(\mathbf{a})$ depends on $\mathbf{a} = \{a_1, a_2, \ldots, a_n\}$. Integration is performed over n variables and is denoted for convenience by the abbreviation

$$\int d\mathbf{a} = \int da_1 \int da_2 \cdots \int da_n. \quad (A2.8)$$

The mean values and mean squared fluctuations are

$$\bar{a}_j = \int d\mathbf{a}\, a_j \rho(\mathbf{a})$$
$$M_{jk} = \int d\mathbf{a}(a_j - \bar{a}_j)(a_k - \bar{a}_k) \rho(\mathbf{a}). \quad (A2.9)$$

The generating function is the average of $\exp(i\Sigma \xi_j a_j)$ and has the form

$$G(\xi) = \exp\left(i \sum_j \xi_j \bar{a}_j - \frac{1}{2} \sum_j \sum_k \xi_j M_{jk} \xi_k \right). \quad (A2.10)$$

The distribution function itself has the form

$$\rho(\mathbf{a}) = \frac{1}{\sqrt{\det(2\pi \mathbf{M})}} \exp\left(-\frac{1}{2} \sum_j \sum_k (a_j - \bar{a}_j)(\mathbf{M}^{-1})_{jk} (a_k - \bar{a}_k) \right) \quad (A2.11)$$

and involves both the inverse of $\mathbf{M}$ and its determinant. The matrix $\mathbf{M}$ must be positive definite; otherwise, there are linear relationships between members of the set $\mathbf{A}$. The normalization factor can be verified by using an orthogonal transformation to a new set of variables, $b_j = \Sigma_k T_{jk} a_k$, and by finding that matrix $\mathbf{T}$ which diagonalizes $\mathbf{M}$. Then, in this new representation, all of the integrals factor, and the determinant appears as a product of the eigenvalues of $\mathbf{M}$. But the determinant is invariant to an orthogonal transformation. So it is the same determinant that appears in the original representation.

The multivariate Gaussian or normal distribution has two important properties. First, if we integrate over any subset of the starting set $\mathbf{a}$, the remaining variables still have a Gaussian distribution. This is most easily seen using the distribution function $\rho_n(\mathbf{a})$ and the generating function $G_n(\xi)$ for an n-variable set. Integration over the last member a_n leads to ρ_{n-1} and G_{n-1}. But this integration corresponds to setting ξ_n equal to zero in G_n; $G_n(\xi_1, \xi_2, \ldots, \xi_{n-1}, 0) = G_{n-1}(\xi_1, \xi_2, \ldots, \xi_{n-1})$. The resulting generating function is still the exponential of a quadratic form, and so the resulting distribution function is still Gaussian.

The other important property is that any linear combination of Gaussian random variables is itself a Gaussian random variable. To see this, use some appropriate transformation **T** such that b_1 is the desired linear combination of the as. Then integrate out the remaining b_2, $b_3, \ldots, b_n$ to get the Gaussian distribution of b_1.

The properties of multivariate Gaussian random variables that have just been presented are helpful in understanding what is meant by "Gaussian random noise." Replace the discrete index j by the continuous index t (or time). Replace sums over j by integrals over t. The mean value of the noise corresponds to $\bar{a}$ and vanishes. The second moment of the noise corresponds to the matrix **M**. If the noise is white (or delta-function correlated), this means that the matrix **M** is diagonal.

A useful device for calculating averages with the Gaussian distribution is based on the identity

$$(a_j - \bar{a}_j)\rho(\mathbf{a}) = -\sum_k M_{jk} \frac{\partial}{\partial a_k} \rho(\mathbf{a}). \qquad (A2.12)$$

Then the average of $(a_j - \bar{a}_j)F(\mathbf{a})$, where **F** is an arbitrary function of **a**, is given by

$$\langle (a_j - \bar{a}_j)F(\mathbf{a}) \rangle = \int d\mathbf{a} (a_j - \bar{a}_j)\rho(\mathbf{a})F(\mathbf{a}), \qquad (A2.13)$$

and by partial integration, using the above identity, we find

$$\langle (a - \bar{a}_j)F(\mathbf{a}) \rangle = \sum_k M_{jk} \left\langle \frac{\partial}{\partial a_k} F(\mathbf{a}) \right\rangle. \qquad (A2.14)$$

This can be used, for example, to work out averages of products. For simplicity of notation, take all $\bar{a}_j = 0$. Then, by applying this formula to the calculation of the fourth moment, we find

$$\langle a_j a_k a_m a_n \rangle = M_{jk} \langle a_m a_n \rangle + M_{jm} \langle a_k a_n \rangle + M_{jn} \langle a_k a_m \rangle, \qquad (A2.15)$$

and for the sixth moment, we find

$$\langle a_j a_k a_m a_n a_p a_q \rangle = M_{jk} \langle a_m a_n a_p a_q \rangle + M_{jm} \langle a_k a_n a_p a_q \rangle + \\ M_{jn} \langle a_k a_m a_p a_q \rangle + M_{jp} \langle a_k a_m a_n a_q \rangle + M_{jq} \langle a_k a_m a_n a_p \rangle, \qquad (A2.16)$$

and so on. In this way, any average of any product of as can be reduced to a sum of products of Ms by successive pairing of indices.

A useful application is to Gaussian white noise. Consider, for example, scalar noise with the second moment

$$\langle F(t_1)F(t_2)\rangle = 2B\delta(t_1-t_2). \quad (A2.17)$$

Then one finds that the fourth moment of the noise is

$$\langle F(t_1)F(t_2)F(t_3)F(t_4)\rangle$$
$$= 4B^2\delta(t_1-t_2)\delta(t_3-t_4) + 4B^2\delta(t_1-t_3)\delta(t_2-t_4)$$
$$+ 4B^2\delta(t_1-t_4)\delta(t_2-t_3). \quad (A2.18)$$

Appendix 3

Laplace Transforms

Many of the dynamical problems encountered in nonequilibrium statistical mechanics are most easily handled using Laplace transforms. This appendix gives a short summary of essential facts about Laplace transforms.

The *definition* is

$$\hat{f}(z) = \int_0^\infty dt\, e^{-zt} f(t) = \mathcal{L}\{f(t), z\}. \quad (A3.1)$$

The carat $\wedge$ is mostly used to denote the transform; the script $\mathcal{L}$ is also used. The transform of a *time derivative* is

$$\hat{\dot{f}}(z) = \mathcal{L}\left\{\frac{df(t)}{dt}, z\right\} = \int_0^\infty dt\, e^{-zt} \frac{df(t)}{dt}$$
$$= \int_0^\infty dt\, \frac{d}{dt}(e^{-zt}f(t)) - \int_0^\infty dt\left(\frac{d}{dt}e^{-zt}\right)f(t) = z\hat{f}(z) - f(0). \quad (A3.2)$$

The transform of the *second time derivative* is done by iterating:

$$\hat{\ddot{f}}(z) = \mathcal{L}\left\{\frac{d^2 f(t)}{dt^2}, z\right\} = z\mathcal{L}\left\{\frac{df(t)}{dt}, z\right\} - \dot{f}(0)$$
$$= z^2\hat{f}(z) - zf(0) - \dot{f}(0). \quad (A3.3)$$

The transform of an *integral* is

$$\mathcal{L}\left\{\int_0^t ds\, f(s), z\right\} = \int_0^\infty dt\, e^{-zt} \int_0^t ds\, f(s) = \int_0^\infty ds \int_s^\infty dt\, e^{-zt} f(s) \quad (A3.4)$$

or, by exchanging the order of integration,

$$\mathcal{L}\left\{\int_0^t ds f(s), z\right\} = \int_0^\infty ds \int_s^\infty dt\, e^{-z(t-s)} e^{-zs} f(s)$$
$$= \int_0^\infty ds \frac{1}{z} e^{-zs} f(s) = \frac{1}{z} \hat{f}(z). \quad (A3.5)$$

The transform of a *convolution* of two functions f and g is the product of the transforms of the two functions:

$$\mathcal{L}\left\{\int_0^t ds f(s)g(t-s), z\right\} = \mathcal{L}\left\{\int_0^t ds f(t-s)g(s), z\right\}$$
$$= \int_0^\infty dt\, e^{-zt} \int_0^t ds f(s)g(t-s) = \int_0^\infty ds \int_s^\infty dt\, e^{-zt} f(s)g(t-s)$$
$$= \int_0^\infty ds\, e^{-zs} f(s) \int_s^\infty dt\, e^{-z(t-s)} g(t-s) = \hat{f}(z)\hat{g}(z). \quad (A3.6)$$

The transforms of some familiar *functions* are

$$\mathcal{L}\{e^{-at}, z\} = \frac{1}{z+a} \quad (A3.7)$$

$$\mathcal{L}\{e^{-at} f(t), z\} = \hat{f}(z+a) \quad (A3.8)$$

$$\mathcal{L}\{t^n, z\} = \frac{n!}{z^{n+1}} \quad (A3.9)$$

$$\mathcal{L}\{\cos \omega t, z\} = \frac{z}{z^2 + \omega^2} \quad (A3.10)$$

$$\mathcal{L}\{\sin \omega t, z\} = \frac{\omega}{z^2 + \omega^2}. \quad (A3.11)$$

Inverting a Laplace transform is generally harder. The easiest way is to use tables of transforms and inverse transforms. As a last resort, one can write the inverse transform as a contour integral in the complex z plane,

$$\mathcal{L}^{-1}\{\hat{f}(z), t\} = \frac{1}{2\pi i} \int_{c-i\infty}^{c+i\infty} dz\, e^{zt}\, \hat{f}(z), \quad (A3.12)$$

where the contour is a straight line parallel to the y axis, located to the right of all singularities of $\hat{f}(z)$ in the z plane. Then the various devices for evaluating contour integrals of complex variables can be tried.

A numerical algorithm for inverting Laplace transforms that often seems to work remarkably well is due to H. Stehfest (*Comm. ACM* 13 (1970): 47–49). It fails when $f(t)$ has discontinuities, sharp peaks, or rapid oscillations. Since time-correlation functions generally do not suffer from these pathologies, the Stehfest algorithm is usually worth trying.

Appendix 4

Continued Fractions

Continued fractions are a useful tool in nonequilibrium statistical mechanics, yet they are seldom mentioned in mathematics courses for scientists. Here is a summary of some essential facts about continued fractions. First, there are two common ways to represent them,

$$f = b_0 + \cfrac{a_1}{b_1 + \cfrac{a_2}{b_2 + \cfrac{a_3}{b_3 + \cdots}}} \tag{A4.1}$$

$$= b_0 + \frac{a_1}{b_1 +} \frac{a_2}{b_2 +} \frac{a_3}{b_3 +} \cdots. \tag{A4.2}$$

The first form is a more graphic representation, and the second is more economical of space. If any a_{n+1} vanishes, the continued fraction terminates at the nth level. Then, by multiplying out, it can be written as a single fraction with a numerator N_n and a denominator D_n,

$$f_n = b_0 + \frac{a_1}{b_1 +} \cdots \frac{a_n}{b_n} = \frac{N_n}{D_n}. \tag{A4.3}$$

For example, truncation at $n = 2$ gives

$$b_0 + \frac{a_1}{b_1 +} \frac{a_2}{b_2} = \frac{b_0 b_1 b_2 + b_0 b_2 + a_1 b_2}{b_1 b_2 + a_2}. \tag{A4.4}$$

An infinite continued fraction is said to converge if this ratio has a limit as n goes to infinity.

An interesting special case is where all the $a_n = 1$ and all the b_n are integers. Then the continued fraction is a representation of an irrational number between 0 and 1. (If the continued fraction terminates, clearly it represents a rational number in this interval.) Another special case is where the continued fraction is periodic; for example,

$$f = \frac{a}{b +} \frac{a}{b +} \frac{a}{b +} \cdots = \frac{a}{b + f}. \tag{A4.5}$$

Note the appearance of f in the denominator. This is a quadratic equation determining f; generally, the numerical value of a periodic continued fraction is an algebraic number.

A practical procedure for computing continued fractions numerically is based on matrix multiplication:

$$f_n = \frac{N_n}{D_n}, \quad \begin{pmatrix} N_n \\ D_n \end{pmatrix} = \begin{pmatrix} N_{n-1} & N_{n-2} \\ D_{n-1} & D_{n-1} \end{pmatrix} \cdot \begin{pmatrix} b_n \\ a_n \end{pmatrix}, \tag{A4.6}$$

which is initialized by $N_{-1} = 1$, $N_0 = b_0$, $D_{-1} = 0$, $D_0 = 1$.

A common application of continued fractions is to solve the vector-matrix equation

$$\frac{d\mathbf{x}}{dt} = \mathbf{L} \cdot \mathbf{x}, \quad x_m(0) = \delta_{m0}, \tag{A4.7}$$

where $\mathbf{L}$ is a symmetric tridiagonal matrix,

$$\mathbf{L} = \begin{pmatrix} a_0 & b_1 & 0 & 0 & \cdots \\ b_1 & a_1 & b_2 & 0 & \cdots \\ 0 & b_2 & a_2 & b_3 & \cdots \\ 0 & 0 & b_3 & a_3 & \cdots \\ \vdots & \vdots & \vdots & \vdots & \ddots \end{pmatrix}. \tag{A4.8}$$

The equation is solved by Laplace transforms; the transform variable is z. Then the equations of motion, for the particular initial condition chosen, are

$$\begin{aligned} (z + a_0)x_0 + b_1 x_1 &= 1 \\ b_1 x_0 + (z + a_1)x_1 + b_2 x_2 &= 0 \\ b_2 x_1 + (z + a_2)x_2 + b_3 x_3 &= 0 \\ b_j x_{j-1} + (z + a_j)x_j + b_{j+1} x_{j+1} &= 0. \end{aligned} \tag{A4.9}$$

The first of these can be rearranged, on dividing by x_0, to

$$x_0 = \frac{1}{z + a_0 + b_1 \dfrac{x_1}{x_0}}. \tag{A4.10}$$

All of the other equations have the same general form because they have no initial value term,

$$b_j \frac{x_{j-1}}{x_j} + (z + a_j) + b_{j+1} \frac{x_{j+1}}{x_j} = 0. \tag{A4.11}$$

This provides a recursion relation connecting ratios of successive terms,

$$\frac{x_j}{x_{j-1}} = -\frac{b_j}{z + a_j + b_{j+1} \dfrac{x_{j+1}}{x_j}}, \tag{A4.12}$$

and application of this recursion produces the continued fraction

$$x_0 = \cfrac{1}{z+a_0 - \cfrac{b_1^2}{z+a_1 - \cfrac{b_2^2}{z+a_2 - \cdots}}} \qquad (A4.13)$$

as a useful representation of the solution of the vector-matrix equation.

Appendix 5

Phenomenological Transport Equations

One of the goals of nonequilibrium statistical mechanics is to provide a molecular basis for the phenomenological transport equations. This appendix gives a brief summary of these equations.

The *diffusion equation* is the simplest of the phenomenological transport equations. It determines the time t and space $\mathbf{r}$ dependence of the concentration $C(\mathbf{r}, t)$ of a diffusing species. First, there is a conservation law,

$$\frac{\partial}{\partial t} C(\mathbf{r}, t) = -\nabla \cdot \mathbf{J}(\mathbf{r}, t), \qquad (A5.1)$$

where $\mathbf{J}(\mathbf{r}, t)$ is a flux. All conservation laws have this form; the integral over the entire confining volume is

$$\frac{\partial}{\partial t} \int_V d\mathbf{r}\, C(\mathbf{r}, t) = -\int_V d\mathbf{r}\, \nabla \cdot \mathbf{J}(\mathbf{r}, t) = -\oint_S d\mathbf{n} \cdot \mathbf{J}(\mathbf{r}, t) = 0 \qquad (A5.2)$$

The integral over the volume is converted to an integral over the bounding surface S of the volume ($d\mathbf{n}$ is a surface area times a unit normal directed out of the volume). But if the diffusing species cannot leave the volume, or is conserved, the flux must vanish on the bounding surface. The total concentration is constant in time.

The flux is given by Fick's law,

$$\mathbf{J}(\mathbf{r}, t) = -D\nabla C(\mathbf{r}, t), \qquad (A5.3)$$

where D is the diffusion coefficient. When these two equations are combined, we obtain the familiar diffusion equation,

$$\frac{\partial}{\partial t} C(\mathbf{r}, t) = D\nabla^2 C(\mathbf{r}, t). \qquad (A5.4)$$

The *hydrodynamic equations* also appear first as conservation laws. The mass density of the fluid is $\rho(\mathbf{r}, t)$, and the local velocity of the fluid is $\mathbf{v}(\mathbf{r}, t)$. The product $\rho\mathbf{v}$ is the local momentum density of the fluid. Then (for simplicity of notation, the space and time dependence are left implicit),

$$\frac{\partial}{\partial t}\rho = -\nabla \cdot \rho \mathbf{v} \tag{A5.5}$$

expresses conservation of mass. The next (actually the second, third, and fourth) equation is a statement of conservation of momentum,

$$\frac{\partial}{\partial t}\rho \mathbf{v} = -\nabla \cdot \mathbf{v}\rho \mathbf{v} - \nabla \cdot \sigma. \tag{A5.6}$$

The first term on the right is the flux of momentum. The quantity σ in the second term is called the stress tensor. It is introduced so that momentum is in fact conserved. Part of the stress tensor is the local equilibrium pressure $P(\mathbf{r}, t)$ of the fluid, and the remainder $\sigma'(\mathbf{r}, t)$ involves deviations from equilibrium,

$$\sigma = P\mathbf{1} - \sigma', \tag{A5.7}$$

where **1** is the unit tensor. When deviations from equilibrium are small, the remainder has the Navier-Stokes form,

$$\sigma' = \eta \left[\nabla \mathbf{v} + (\nabla \mathbf{v})^\dagger - \frac{2}{3}\nabla \cdot \mathbf{v}\mathbf{1} \right] + \eta_V \nabla \cdot \mathbf{v}\mathbf{1}. \tag{A5.8}$$

In this equation, η is the coefficient of shear viscosity, and η_V is the coefficient of volume viscosity. $\nabla \mathbf{v}$ is the velocity gradient tensor. Often one writes $(\nabla \mathbf{v})^\dagger$ as $\mathbf{v}\nabla$.

The final hydrodynamic equation is a statement of energy conservation. The total energy density is the sum of the internal energy density E and the kinetic energy of the bulk flow of the fluid, $E + \rho v^2/2$. The internal energy depends on the local temperature and mass density or on any other pair of thermodynamic quantities (pressure and entropy are often used). The equation of conservation of energy is

$$\frac{\partial}{\partial t}\left[\frac{1}{2}\rho v^2 + E\right] = -\nabla \cdot \left[\mathbf{v}\left(\frac{1}{2}\rho v^2 + E\right)\right] - \nabla \cdot (\mathbf{v} \cdot \sigma) - \nabla \cdot \mathbf{q}. \tag{A5.9}$$

The first term contains the flux of total energy. The second term contains energy production by work done against internal stresses (including pressure). The final term contains the heat flux **q**; when the deviation from equilibrium is small, this is given by the Fourier heat law,

$$\mathbf{q} = -\kappa \nabla T, \tag{A5.10}$$

where κ is the coefficient of thermal conductivity.

The preceding five equations, augmented by the Navier-Stokes stress tensor and the Fourier heat law, describe an enormous range of physical phenomena. Their solution, even in very simple circumstances, can lead to great mathematical difficulty. Our concern here, however, is with their molecular foundation rather than with their solution.

References

These references are mainly to a few historically important papers, and to a few little-known but useful papers.

Alder, B. J. and Wainwright, T. E. 1968 *J. Phys. Soc. Japan* 26, 267
Bethe, H. and Teller, E. 1941 *Ballistic Research Laboratory Report* X117
Bhatnagar, P. L., Gross, E. P., and Krook, M. 1954 *Phys. Rev.* 94, 511
Grad, H. 1949 *Commun. Pure Appl. Math.* 2, 331
Hynes, J. and Deutch, J. 1975, in *Physical Chemistry, an Advanced Treatise*, edited by H. Eyring, D. Henderson, and W. Jost, vol. XI, Academic Press, New York.
Kadanoff, L. and Swift, J. 1968 *Phys. Rev.* 166, 89
Kramers, H. A. 1940 *Physica* 7, 284
Kubo, R. 1957 *J. Phys. Soc. Japan* 12, 570
Landau, L. and Teller, E. 1936 *Physik. Z. Sowjetunion* 10, 34
Marcus, R. A. 1960 *Discuss. Faraday Soc.* 29, 21
Montroll, E. W. and Shuler, K. E. 1957 *J. Chem. Phys.* 26, 454
Mori, H. 1965 *Prog. Theor. Phys.* 33, 423
Nakajima, S. 1958 *Prog. Theor. Phys.* 20, 948
Nordholm, S. and Zwanzig, R. 1975 *J. Stat. Phys.* 13, 347
Onsager, L. 1931 *Physical Review* 37, 405
Rubin, R. J. 1960 *J. Math. Phys.* 1, 309
Van Hove, L. 1955 *Physica* 21, 517
Wigner, E. P. 1932 *Z. Phys. Chem. Abt. B* 19, 203
Zhou, H.-X. and Zwanzig, R. 1991 *J. Chem. Phys.* 94, 6147
Zwanzig, R. 1960 *J. Chem. Phys.* 33, 1338
Zwanzig, R. 1961 *Phys. Rev.* 124, 983

Index

Adjoint operators, Fokker–
Planck equations 42–43

BGK equation
  approximation to Boltzmann
    equation 93
  defining local equilibrium
    distribution 94
  H-theorem derivation 95–96
  hydrodynamics 96–100
  ratio of thermal conductivity
    to viscosity 100
Bloch equations
  approximate equation for
    average spin matrices 118
  average magnetization 117
  density matrix 118
  derivation 115–121
  energy absorption 120–121
  memory kernel 118–119
  time correlation function
    118–119
Brownian motion
  harmonic oscillator heat bath
    21–24
  heavy mass in harmonic lattice
    24–28
  kinetics of first-order
    isomerization reaction
    14–15
  Langevin equation for
    harmonic oscillator 17–18
  mean first passage time 77–78
  molecular dipole in periodic
    potential 18
  nonlinear equations for slow
    variables 183–184
Brownian particle
  force by interaction with
    medium 4
  friction coefficient by Stokes'
    law 4
  mean squared displacement
    11–12
  self-diffusion coefficient 12
  velocity correlation function
    10–11
Brownian rotator, orientational
  time correlation function
    45–47

INDEX

Chemical kinetics
  bimolecular reaction 64
  concentration variables 65
  Fokker–Planck equations 65
  rate equation 66
  use of master equations 64–66
Chemical reaction rates, transition state theory 67–72
Collisions
  hydrodynamics 97
  rate of occurrence for BGK equation 95
  rotational diffusion 85
Continued fractions
  common ways to represent 205
  expansion 91
  evaluating by truncation 88
  recursion relation 206–207
Convolution, Laplace transforms 204
Correlation function, spectral density 139
Cumulants
  definition 189
  equation for cumulant generating function 190
  perturbation expansion for equation of motion for 190–192

Delta function
  occurrence in Golden Rule formula 52
  use in derivation of identities 139–141
Density matrix
  equilibrium 103
  partitioning into diagonal and off-diagonal parts 125
  quantum Liouville equation 106, 133
  two-level system in heat bath 110–111, 116–117

Dephasing, Hamiltonian as elementary model 110
Differential equations
  first-order linear 198–200
  second-order 199
Diffusion
  one-dimensional 8–9
  solution of ordinary equation in angle space 88
  three-dimensional 10
  transport equation 207
  velocity correlation function 8–10
  See also Energy diffusion
Dipole-dipole correlation function
  absorption coefficient proportional to spectral density of 12
  Langevin equation for rotational Brownian motion 13
  orientational time correlation function 13–14
  two-level system in heat bath 111
Dynamic linear response
  classical mechanics 130–132
  density matrix 133
  frequency dependent response 135–136
  Liouville equation 130–131, 133–134
  phase space distribution function 133
  Poisson bracket 130–131
  quantum commutator 133
  quantum mechanics 133–135
  time-dependent analog of static susceptibility 132
  total response a sum of individual responses 132

Electron transfer kinetics
  heat bath Hamiltonian 107–108

INDEX   215

Marcus's reorganization energy 110
polarization of environment 106–107
quantum analog of Kramers problem 106
rate of transition between states by Golden Rule formula 108
spin-boson Hamiltonian 108
rate of energy dissipation 139
Energy absorption rate, optical absorption coefficient 53–55
Energy diffusion
energy diffusion coefficient 80
energy diffusion equation 81
escape rate as function of friction 82f
mean first passage time 82
rate of escape from potential well over barrier 81

First-order linear differential equations 198–200
First passage times
adjoint equation 75–76
derivation 74–76
distribution 74
Kramers problem 76–78
mean 75
Fluctuation-dissipation theorem
analog of, for Langevin equation 16–17
arbitrary distinction between systematic behavior and noise 24
balance between friction and noise 6–7
correlation function of noise 161
harmonic oscillator example 23
non-Markovian version 20
non-Markovian version in matrix form 21
requiring symmetry and antisymmetry 17
steady-state solution of Fokker–Planck equation 39
two-variable Brownian motion of particle moving in potential 39
Fluctuation-renormalization
bare transport coefficient 192
change in memory functions 160–161
nonlinear Langevin equations 188
Fokker–Planck equations
averages and adjoint operators 42–43
choice of initial conditions 197
derivation 36–39
derivation of nonlinear 177–180
Green's function in linear case 43–44, 187–188
Heisenberg approach 42–43
long time steady-state solution of arbitrary 39
noise-averaged distribution function 38, 175
nonlinear 180–181
properties 41–42
Schrodinger approach 42
slow variables 183, 187
Smoluchowski equation 40–41
substitution of Smoluchowski equation leading to Schrodinger-like equation 41
two-variable Brownian motion of particle moving in potential 39–40
Frequency dependent magnetic susceptibility, linear response theory 137–138
Frequency dependent response, linear response in quantum mechanics 135–136

INDEX

Gaussian random variables
  application to Gaussian white noise 202–203
  distribution function 201
  linear combination of 202
  mean values and mean squared fluctuations 201
  moment generating function 200–201
  noise 23
  probability distribution 200
  properties of multivariate 201–202
Golden Rule
  definition 48
  derivation 48–51
  energy absorption equation 142
  flaws of standard treatment 52–53
  short time behavior 53
  transition states in Pauli master equation 124
  uses in electron transfer kinetics 108–109
Gram–Schmidt process, orthonormalizing vectors in Hilbert space 146

Harmonic lattice
  instructive model of Brownian motion 24
  Laplace transforms for solving equations of motion 25
  recurrence paradox 195–196
  reversibility paradox 195
  transform of normalized velocity correlation function 26
  velocity correlation function 24–25, 27–28
Harmonic oscillator
  Brownian motion in, heat bath 21–24
  heat bath master equation 60–61
  Langevin equation 22
  Langevin equation for Brownian motion 17–18
  memory function 22–23
  noise 23, 24
Heat bath
  Bloch equations, two-level system 115–121
  Brownian motion in harmonic oscillator 21–24
  dephasing, two-level system 110–115
  See also Two-level system
Heat bath master equation
  application 59–61
  derivation 126
  Golden Rule transition rates 58
  harmonic oscillator 60–61
  microcanonical character 59
  principle of detailed balance 59
Heisenberg equations of motion, electron transfer kinetics 107
Hilbert space
  basis of mode-coupling theory 170
  expansion of velocity correlation function 171
  matrix form of Liouville equation 144–146
  projection operators 143–149
  quantum mechanics 101
  subspace of relevant variables 148–149

Inverting Laplace transforms 204
Ion mobility, linear response theory 136–137
Isomerization reaction, kinetics of first-order 14–15

Kinetic models
  BGK equation and H-theorem 93–96
  BGK equation and hydrodynamics 96–100
  Boltzmann equation 83–84
  collision integral 84
  orientational time correlation function 89–92
  rotational diffusion 84–88
  rotational relaxation 89–92
Kramers–Kronig relation, frequency dependent response 136
Kramers problem
  Brownian particle escape from potential well 73, 76
  electron transfer reactions as quantum analog of 106
  energy diffusion concept 78–82
  mean first passage time 77–78
  relation to transition state theory rate 78
Kubo transform
  linear response in quantum mechanics 135
  noise in generalized Langevin equation 155

Langevin equations
  Brownian motion of harmonic oscillator 17–18
  choice of initial conditions 197
  derivation for Brownian motion of system with harmonic oscillator heat bath 21–24
  derivation of nonlinear 177–180
  general treatment 15–16
  Markovian and non-Markovian 19–21
  nonlinear 18–19, 180

Langevin equations, generalized
  derivation 149–151
  deriving Fokker–Planck equation 177
  eliminating projection operators 158–159
  identities 157–160
  initial nonequilibrium states 151–155
  linear, for slow variables 165–168
  Mori's procedure 161–162
  noise 151–157
  nonlinear to linear example 160–165
  non-Markovian fluctuation-dissipation theorem 157
  procedure constructing initial nonequilibrium distribution 154–155
Laplace transforms
  continued fractions 206
  convolution 204
  definition 203
  functions 204
  handling time derivatives 87
  integral 203–204
  inverting 204
  orientational time correlation function, numerical inversion 92f
  solving equations of motion of harmonic lattice 25
Linear differential equations, first-order 198–200
Linear response
  applications of theory 136–139
  determining equilibrium 127–129
  dynamic 130–136
  dynamic in classical mechanics 130–132
  dynamic in quantum mechanics 133–135
  energy absorption 141–142

Linear response (*cont.*):
  energy absorption in electric field 138–139
  frequency dependent magnetic susceptibility 137–138
  frequency dependent response 135–136
  identities 139–142
  mobility of ion 136–137
  quantum mechanical version 129–130
  quantum perturbation theory 130
  static 127–130
Linear response theory
  applications 136–139
  initial ensemble 197
  streaming velocity 186
Liouville equation
  dynamical variables 33–35
  evolution of dynamical variable 34
  formal operator solution 32
  Liouville operator 32–33
  matrix form 144–146
  partitioning 146–147
Long time tails
  deriving, stress correlation function 173
  mode-coupling theory 169–170

Magnetic susceptibility, linear response theory 137–138
Master equations
  abstract 61–63
  chemical kinetics 64–66
  derivation of quantum mechanical or Pauli 121–124
  Golden Rule transition rates of heat bath 58
  harmonic oscillator illustration of heat bath 59–61
  heat bath 57–59
  matrix or operator equation 62
  Pauli 56–57
  random walks 63–64
  use of operator methods for averages 62
Memory kernel
  derivation of nonlinear equations 179–180
  derivation of Pauli master equation 123, 126
  eliminating projection operators 158–159
  linear Langevin equations for slow variables 165
  non-Markovian fluctuation-dissipation theorem 157
  two-level system in heat bath 118–119
Mobility of ion, linear response theory 136–137
Mode-coupling theory
  deriving long time tail of stress correlation function 173
  Hilbert space picture of dynamics 170
  long time tails 169–170
  product of two slow variables 171
  self-diffusion example 170–173
Mori Langevin equation
  differences between, and exact Langevin 160
  memory function 164–165
  noise 165
  nonlinear to linear example 160–165
  slow variables 166

Nernst–Planck equation, treatment of electrolyte solutions 160
Noise
  application of Gaussian random variables to white 202–203
  arbitrary distinction from systematic behavior 24

averaged, in Langevin
    equation 155–156
generalized Langevin
    equations 151–157
nonlinear Langevin equation
    184–185
Nonlinear Langevin and
    Fokker–Planck equations
    derivation 177–181
    illustration of nonlinear system
        interacting with harmonic
        oscillator heat bath 183–184
    memory kernel 179–180
    noise and initial states
        184–185
    reduced distribution functions
        175–177
    slow variables 181–183

Optical absorption coefficient
    classical time correlation
        function of total electric
        dipole moment 56
    focus on rate of energy
        absorption 53–55
    frequency dependence 12
    frequency dependence by time
        correlation function 53
    theory of optical absorption
        55–56
Orientational time correlation
    function
    approaching ideal rotator limit
        91
    deriving exact expression
        89–92
    exponential decay 92
    Laplace transform of
        correlation function 90–91
    results of numerical inversion
        of Laplace transform 92f

Pauli master equation
    derivation 121–124
    Golden Rule 124

memory kernels 123, 126
microcanonical character 57
projection operator method
    124–126
Probability distribution, Gaussian
    random variables 200
Projection operator method
    deriving quantum mechanical
        master equation 124–126
    use in deriving Langevin
        equations 143–144
Projection operators
    derivation of generalized
        Langevin equations 149–151
    deriving Fokker–Planck
        equation 177
    deriving Langevin equations
        178
    eliminating projection 158–159
    Hilbert space 143–149
    identities of generalized
        Langevin equations 157–160
    initial nonequilibrium states of
        noise 151–155
    linear Langevin equations for
        slow variables 165–168
    matrix form of Liouville
        equation 144–146
    Mori's linear versus exact
        Langevin equation 160–165
    noise in generalized Langevin
        equations 151–157
    nonlinear to linear example
        160–165
    non-Markovian fluctuation-
        dissipation theorem 157
    partitioning 146–147
    projecting 147–148
    quantum mechanics 157
    subspace of relevant variables
        148–149
    time correlation functions 158

Quantum Liouville operator
    anti-self-adjoint 106

Quantum Liouville operator (*cont.*):
  Heisenberg representation 105
  modifications for time-dependent Hamiltonian 105
  tetradic representation 105
  time-dependent density matrix 106
  time-dependent Heisenberg operator 104
  von Neumann equation 106
Quantum statistical mechanics
  energy eigenvalues and eigenfunctions 101–102
  equilibrium density matrix 103
  Hamiltonian operator 101
  quantum Liouville operator 104–106
  thermal equilibrium average 103
  trace of matrix 103
  transformation matrix connecting two sets of states 102–103

Random walks
  application of master equations 63–64
  integral as modified Bessel function 64
Reaction rates
  example of ideal gas model system 72–73
  first passage times 74–76
  Kramers problem 76–78
  Kramers problem and energy diffusion 78–82
  transition state theory 67–72
Recurrence paradox
  example of one-dimensional harmonic lattice 195–196
  objection to Boltzmann's H-theorem 194
  recurrence time 196

Reversibility paradox
  example of one-dimensional harmonic lattice 195
  objection to Boltzmann's H-theorem 193–194
  one-way irreversible character appearing to violate time-reversal symmetry 193–194
Rotational diffusion
  derivation of equation 84–88
  evaluating continued fractions by truncation 88
  kinetic model 84–88

Shear viscosity
  coefficient 99
  deriving formula 99
  ratio to thermal conductivity 100
  time correlation function determining 168
Slow variables
  diffusion coefficient 182, 186–187
  equilibrium distribution 185–187
  Fokker–Planck equation 183
  Langevin equation 183
  streaming velocity, Langevin equation 185–186
Smoluchowski equation
  first passage times 75
  Fokker–Planck equation 40–41
  substitution leading to Schrodinger-like equation 41
Spectral density
  absorption coefficient proportional to, of dipole-dipole time correlation function 12
  correlation function 139
  Fourier transform of time correlation function 8

Spin-boson Hamiltonian,
  electron transfer kinetics
  108
Static linear response
  determining equilibrium
  127–129
  electric dipole moment 128
  Kubo transform 129–130
  quantum mechanical version
  129–130
  quantum perturbation theory
  130
  unperturbed and perturbed
  distribution function 128
  unperturbed and perturbed
  partition function 128
Stokes–Einstein formula, friction
  coefficient 12
Stress tensor
  hydrodynamics 97
  Navier–Stokes form of viscous
  99
Stress time correlation function
  deriving long tail time 173
  long time tails 170
Superoperator, quantum
  commutator 104

Taylor's series expansion, time-
  dependent dynamical
  variable 34
Thermal conductivity
  deriving formula 99–100
  Fourier's heat law 100
  heat current 99–100
  mode-coupling theory 169
  ratio to viscosity 100
Time correlation functions
  combining Bloch equations 120
  derivation for orientational
  89–92
  determining properties of
  systems out of equilibrium 7
  dipole-dipole correlation
  function 12–14

equilibrium fluctuations in
  particle number 15
fluctuating magnetic field 60
fluctuation, nonlinear for slow
  variables 184
frequency dependence of
  optical absorption
  coefficient 53
generalized Langevin
  equations 158
integral over angular velocities
  only 89
Langevin equation 7–10
Liouville operator notation
  35–36
mean squared displacement
  11–12
orientational, of planar
  Brownian rotator 45–47
spectral density 8
statistical behavior of time-
  dependent quantity 7–8
Stehfest algorithm 92
transition state theory 72
two-level system in heat bath
  118–119
using partition functions 8
velocity correlation function
  8–11
Time derivative, Laplace
  transforms 203
Transition state theory
  alternative form using
  quantum mechanical
  partition function 70–72
  approximation 68
  escape rate 73
  example of ideal gas escape
  from two-dimensional
  region 72–73
  flux density 68
  Hamiltonian 68–69
  model system 73f
  partition function of transition
  state 70–71

222 INDEX

Transition state theory (*cont.*):
  phase space distribution function 67
  probability of being in region 68
  rate constant 69–70
  rate constant as ratio of two partition functions 70
  rate equations 69–70
  rates of chemical reactions 67
  relation of Kramers rate to, rate 78
  time correlation function 72
  transitions between regions by simple first-order kinetics 71
Transport equations
  diffusion equation 207
  hydrodynamic equations 207–208
  phenomenological 207–209
Two-level system
  approximate methods 113–115
  caution for Markovian approximations 115
  density matrix 110–111
  dipole time correlation function 111–112
  equation of motion for density matrix 111
  exact solution procedure 112–113
  heat bath, Bloch equations 115–121
  heat bath, dephasing 110–115
  spectral line shape by dipole-dipole line correlation function 111

Velocity correlation function
  Brownian particle 10–11
  connection with self-diffusion coefficient 8–10
  fluctuation-dissipation theorem 11
  heavy mass in harmonic lattice 24–28
  Hilbert space expansion 171
  ion mobility 137
  long time tails 169–170
  mode-coupling theory 171–173
  random behavior at long times 196*f*
von Neumann equation, quantum Liouville equation 106

CPSIA information can be obtained at www.ICGtesting.com
Printed in the USA
BVOW020712140113

310477BV00004B/12/A